The Muscular System

Other titles in
Human Body Systems

The Circulatory System by Leslie Mertz

The Digestive System by Michael Windelspecht

The Endocrine System by Stephanie Watson and Kelli Miller

The Lymphatic System by Julie McDowell and Michael Windelspecht

The Nervous System and Sense Organs by Julie McDowell

The Reproductive System by Kathryn H. Hollen

The Respiratory System by David Petechuk

The Skeletal System by Evelyn Kelly

The Urinary System by Stephanie Watson

The Muscular System

Amy Adams

HUMAN BODY SYSTEMS
Michael Windelspecht, Series Editor

Greenwood Press
Westport, Connecticut • London

Library of Congress Cataloging-in-Publication Data

The muscular system / Amy Adams.
p. cm.—(Human body systems)
Includes bibliographical references and index.
ISBN 0–313–32403–4 (alk. paper)
1. Muscles—Physiology. 2. Muscles—Diseases. 3. Muscles—Research—History.
I. Adams, Amy. II. Human body systems.
QP321.M8965 2004
612.7'4—dc22 2004047595

British Library Cataloguing in Publication Data is available.

Library of Congress Catalog Card Number: 2004047595
ISBN: 0–313–32403–4

First published in 2004

Greenwood Press, 88 Post Road West, Westport, CT 06881
An imprint of Greenwood Publishing Group, Inc.
www.greenwood.com

Printed in the United States of America

The paper used in this book complies with the Permanent Paper Standard issued by the National Information Standards Organization (Z39.48–1984).

10 9 8 7 6 5 4 3 2

Illustrations, unless otherwise credited, are by Sandy Windelspecht.

The *Human Body Systems* series is a reference, not a medical or diagnostic manual. No portion of this series is intended to supplement or substitute medical attention and advice. Readers are advised to consult a physician before making decisions related to their diagnosis or treatment.

Contents

Color photos follow p. 134.

Series Foreword

Human Body Systems is a ten-volume series that explores the physiology, history, and diseases of the major organ systems of humans. An organ system is defined as a group of organs that physiologically function together to conduct an activity for the body. In this series we identify ten major functions. These are listed in Table F.1, along with the name of the organ system responsible for the activity. It is sometimes difficult to specifically define an organ system, because many of our organs have dual functions. For example, the liver interacts with both circulatory and digestive systems, the hypothalamus acts as a junction between the nervous and endocrine systems, and the pancreas has both digestive and endocrine secretions. This complex interaction of organs and tissues in the human body is still not completely understood.

This series is unique in that it provides a one-stop reference source for anyone with an interest in the human body. Whereas other references frequently cover one aspect of human biology, from anatomy and physiology to the prevention of diseases, this series takes a more holistic approach. Each volume not only includes a physiological description of how the system works from the cellular level upward, but also a historical summary of how research on the system has changed since the time of the ancients. This is an important aspect of the series and one that is frequently overlooked in modern textbooks. In order to understand the successes and problems of modern medicine, it is first important to recognize not only the achievements of the past but also the misunderstandings and challenges of the pioneers in medical research.

For example, a visit to any major educational institution reveals large lec-

TABLE F.1. Organ Systems of the Human Body

Organ System	General Function	Examples
Circulatory	Movement of chemicals through the body	Heart
Digestive	Supply of nutrients to the body	Stomach, small intestine
Endocrine	Maintenance of internal environmental conditions	Thyroid
Lymphatic	Immune system, transport, return of fluids	Spleen
Muscular	Movement	Cardiac muscle, skeletal muscle
Nervous	Processing of incoming stimuli and coordinates activity	Brain, spinal cord
Reproductive	Production of offspring	Testes, ovaries
Respiratory	Gas exchange	Lungs
Skeletal	Support, storage of nutrients	Bones, ligaments
Urinary	Removal of waste products	Bladder, kidneys

ture halls, where science instructors present material to the students on the anatomy and physiology of the human body. Sometimes these classes include laboratory sessions, but in the study of human biology, especially for students who are not bound for professional schools in medicine, the student's exposure to human biology typically centers on a two-dimensional graphic. Most educators accept this process as a necessary evil of the educational system, but few recognize that, in fact, the large lecture classroom is the product of a change in Egyptian religious beliefs before the start of the current era. During the decline of the Egyptian empire and the simultaneous rise of the ancient Greek culture, the Egyptian religious organizations began to forbid the dissection of the human body. This had a twofold influence on medicine. First, the ending of human dissections meant that medical professionals required lectures from educators, instead of participation in laboratory-based education, which led to the birth of the lecture hall. Thus the birth of the lecture hall. The second consequence would plague modern medicine for a thousand years. Stripped of their access to human cadavers, researchers studied other "lesser" animals and extrapolated their findings to humans. The practices of the ancient Greeks were passed on over the ages and became the basis for the study of modern medicine. These traditions continue to this day throughout the educational institutions of the world.

The history of human biology parallels the development of modern science. In the seventeenth century, William Harvey's study of blood circulation challenged the long-standing belief of the ancient Greeks that blood was produced in the liver and consumed in the tissues of the body. Harvey's pioneering experimental work had a strong influence on others, and within a century the legacy of the ancient Greeks had collapsed. In the eighteenth century a group of chemists who focused on the chemical reactions of the human body, called the iatrochemists, began to apply chemical laws to human physiology. They were joined by the iatrophysicists, who believed that the human body must operate under the physical laws of the universe. This in turn led to the beginnings of organic chemistry and biochemistry in the nineteenth century, as scientists focused on identifying the building blocks of living cells and the chemical reactions that they utilize in their metabolism.

In the past century, especially in the last three decades, the rapid advances in technology and scientific discovery have tended to separate most sciences from the general public. Yet despite an ongoing trend to leave the majority of the physical sciences to the scientists, interest in the human biology has actually increased among the general population. This is primarily due to medical discoveries that increase not only lifespan but also healthspan, or the number of years that people live disease free. But another important aspect of this trend is the desire among the general public to be able to ask intelligent questions of their physicians and seek additional information on prescribed medications or procedures. In many cases this information serves as a system of checks and balances on the medical profession, ensuring that the patient is kept well informed and aware of the fundamentals regarding the procedure.

This is one of the most remarkable ages in the study of human biology. The recently announced completion of the Human Genome Project is an indication of how far biology has progressed. Barely fifty years ago, scientists were first discovering the structure of DNA. They now are in possession of an entire encyclopedia of human genetic information, and although they are not yet exactly sure what the content reveals, scarcely a week goes by without a researcher announcing a medical discovery that was made possible by the availability of the complete human genetic sequence. Coupled to this are the advances in the development of pharmaceuticals and treatments that were unheard of less than a decade ago.

But these benefits to society do not come without a cost. The terms stem cells, cloning, and gene therapy no longer belong to the realm of science fiction. They represent advances in the sciences that may hold the key to increased longevity. However, in many cases they also produce ethical and moral questions of society: Where do medical researchers obtain the embryonic stem cells for their work? Who will determine if humans can be cloned? What are the risks of transgenic organisms produced by gene therapy? These are just a few of the potential conflicts that face modern society. Only a well-educated general public can intelligently survey the pros and cons of an ethical or moral decision regarding medical science. Armed with information, concerned people can participate in the democratic process of informing their elected officials of their concerns. Science education is an important aspect of citizenship, and thus the need for series such as this to present information to the general public.

This volume covers the biology of the muscular system. Muscles are one aspect of human biology that we are all familiar with. Nearly everyone at some point has experienced muscle fatigue, pain, or injury. Yet the muscles that regulate our movement are only one part of the muscular system. There are hundreds of muscles in the human body, and together they control not only physical movement but also the flow of blood and fluids, the secretion of glands, and the movement of material within the digestive tract. The importance of the muscles is apparent by the physiological consequences of muscular injury. Strains of skeletal muscles can produce severe pain and lack of mobility, weakening of smooth muscles may reduce the movement of materials through the gastrointestinal system, and failure of the cardiac muscles frequently results in death. As we age, we are all plagued in some form by muscle-related ailments and thus have a need for an information source such as this volume dedicated to the muscular system.

The ten volumes of the *Human Body Systems* series are written by professional authors who specialize in the presentation of complex scientific ideas to the general public. Although any book on the human body must include the terminology and jargon of the profession, the authors of this series keep it to a minimum and strive to explain the concepts clearly and concisely. The series is ideal for the public libraries as well as for secondary school and introductory college libraries. In addition, medical professionals or anyone with an interest in human biology would find this series a useful addition to their personal library.

Michael Windelspecht
Blowing Rock, North Carolina

Introduction

One characteristic that unites all animals is the ability to move from place to place. Animals walk, crawl, or swim to find food, avoid prey, reproduce, and live out their lives. All of this movement requires muscles. In simple organisms, a few muscles are enough to control how the animal gets around. But in complex animals such as humans, a complex network of different muscle types helps fine-tune our movements. Large, powerful muscles allow us to walk, while delicate muscles give us the ability to make the detailed movements needed to write or play the piano. In addition to helping us navigate the world, muscles are essential to our internal processes. The heart beats more than 1 billion times in an average human life, and each of those beats is the result of the heart muscle contracting. Other muscles in the body are responsible for moving food through the digestive system, causing air to rush into and out of the lungs, or directing the blood as it flows through the circulatory system. This volume provides a detailed overview of the anatomy, physiology, and medical issues affecting the many muscles in our body.

Humans have three distinctly different types of muscles: Our roughly 600 skeletal muscles allow us to navigate the world; the heart muscle keeps the heart beating; and smooth muscles line our internal organs, digestive tract, and veins. Although people have been interested in the question of how muscles contract since the era of ancient Greek civilization, it wasn't until the invention of the microscope in the mid-1800s that scientists first identified the three different types of muscles. Even then, scientists did not understand how the muscles contracted until 1952, when the newly invented electron microscope revealed the fine details of the muscles. Until that time, scientists had argued about how the proteins within the muscles interact to cause the muscle to contract. Another long-standing question had concerned what gives the muscles a signal to contract. As long ago as the 1500s, early scientists understood that nerves were needed to deliver signals to muscles, but without a detailed understanding of the nervous system and circulatory system scientists studying the muscles could not fully understand how they function.

Now scientists have a detailed understanding of how nerves signal muscles to contract and exactly how that contraction takes place. However, that doesn't mean the field of muscle physiology no longer has important avenues of research. Scientists still tackle questions about how the muscles respond to exercise, how the muscles form in a developing embryo, how the muscles heal after an injury, and what goes wrong in muscular diseases such as muscular dystrophy. The way this research benefits most people is through a better understanding of how to train for athletic events, a better understanding of what nutrition muscles need during exercise, and better treatments for injured muscles. This research is also starting to produce some methods for helping people with muscular dystrophy, although there is still no cure for the debilitating disease.

With all the functions they perform, muscles interact closely with other systems in the body. The blood vessels winding throughout muscles provide food and oxygen while carrying away waste products. Nerves deliver the essential signal that causes the muscles to contract and also play a role in shaping how muscles develop in the growing embryo. Some muscles are critical in order for other body systems to function normally. The digestive system, for example, relies on smooth muscles to push food from one end of the body to the other, and smooth muscles lining the bladder regulate urination. With their widespread functions, muscles are involved in just about every aspect of human physiology.

Other books in this ten-volume series cover the human body's many biological systems, including some that are integral to muscle function. This volume is intended to be a reference for people who are interested in an overview of muscle physiology, research, and diseases. Following a list of interesting facts, the book is divided into three sections. The first four chapters are devoted to the anatomy and physiology of muscles, including information about exercise physiology, how muscles develop, and how muscles generate energy to contract. Chapters 5 and 6 describe how muscles have been studied throughout history and what types of muscular research continue today. This section also describes some of the techniques being used to better understand muscular diseases. The final four chapters focus on medical aspects of muscles, including both muscle injuries and diseases of the three muscle types. Of diseases affecting the skeletal muscles, muscular dystrophy is by far the most common. This volume describes the many types of muscular dystrophy and discusses how doctors diagnose and treat the conditions, but it is not meant to be a diagnostic tool. Only a doctor with up-to-date information about muscular disorders can diagnose a disease and recommend treatment plans. Likewise, this volume is not meant to replace the advice of a doctor or coach for athletic training and injury recovery.

The final pages of the volume include a list of acronyms used throughout the text, a glossary, a list of organizations and Web sites, a bibliography, and an index to help quickly locate sections of interest. Within the text, words in **bold** indicate the first use of a word that is defined in the Glossary. The text also includes references to other volumes in this series where a reader can find more information on topics that are outside the scope of this volume.

INTERESTING FACTS

- The body contains roughly 630 skeletal muscles.
- The skeletal muscles account for roughly 50 percent of the body weight in men, 40 percent of the body weight in women, and 25 percent of a baby's body weight.
- After age 50, people lose about 10 percent of their muscle fibers per decade.
- Resting muscles receive about 20 percent of blood flow.
- During heavy exercise, the muscles receive from 60 to 85 percent of blood flow.
- Five to 10 percent of a person's body weight is heart and smooth muscle.
- A fast-twitch muscle reaches peak contraction in about $\frac{1}{20}$ of a second.
- A slow-twitch muscle reaches peak contraction in about $\frac{1}{10}$ of a second.
- A single motor unit can range from two to three muscle fibers in the larynx to 2,000 fibers in the hamstring.
- Each sarcomere shortens between one-half to one-third its total length during a contraction.
- Muscles use up stored adenosine triphosphate (ATP) in about two seconds.
- Most pain that occurs the day after exercise is the result of eccentric exercise.
- Pain following exercise occurs most often in the region of the muscle farthest from the center of the body.
- Most strains happen at the junction between the muscle and the tendon.
- Muscular dystrophy is the second most common lethal genetic disease for a child to be born with.
- One in 3,300 male babies are born with Duchenne muscular dystrophy.

Anatomy of the Muscular System

People usually think about muscles being used for running, jumping, or lifting heavy objects. But even sitting and reading a book uses muscles throughout the body. Muscles control the eyes moving back and forth across the page, hold the book upright, maintain posture in a chair, and allow the eyes to focus on the page. Other muscles in the body carry on the day-to-day jobs that go unnoticed such as pumping blood, moving food through the digestive tract, and breathing.

There are three basic types of muscles that carry out all movement within the body: **skeletal muscle**, **smooth muscle**, and **cardiac (heart) muscle**. Although each of these three types of muscles shares the ability to contract, they are located in different places in the body, contract at different strengths, and look different under a microscope.

TYPES OF MUSCLES

Smooth Muscle

Smooth muscle lines many of the internal organs. They cause contractions that move food through the intestines, expand and contract the blood vessels to regulate blood supply, and contract to push the baby out of a woman's uterus. Smooth muscles also control the size of the pupil and are found at the base of hair follicles where they produce goose bumps when it's cold. Ideally these goose bumps would raise the hair to trap an insulating layer of air around the body, although few humans have enough hair for goose bumps to serve a practical purpose.

Smooth muscles are also called involuntary muscles because they can't be controlled voluntarily; one can't decide to relax the pupil to let in more light any more than a person can stop the spread of a blush across the face. Instead, smooth muscles contract in response to biological conditions such as food passing through the intestine, bright light shining on the eye, cold temperatures, or embarrassment.

Smooth muscles contract slowly but with great force and can hold a contraction without growing tired. They also shorten more when they contract than other muscles do. Whereas most muscles are designed for quick, precise motions, smooth muscles are designed for long-term squeezing.

Smooth muscles earned their name by being smooth in appearance under the microscope. The individual cells are long and tapered and tend to form into sheets, such as in the lining of the digestive system. As with most cells in the body, each smooth muscle cell has its own nucleus. (See "Muscles across Species" for a discussion of the three muscle types in other species.)

Cardiac Muscle

Cardiac muscle is the muscle that makes up the heart. It is similar to smooth muscle in that (1) each cardiac muscle cell has its own nucleus and (2) the heart muscle cannot be controlled voluntarily. Instead, a region of the brain monitors how much oxygen is in the blood and adjusts how quickly the heart beats accordingly. A runner can't decide to slow the heart rate, nor can a resting person voluntarily speed up how quickly the heart beats.

The cardiac muscle is extremely strong and is unique in that the entire muscle contracts at the same time. Whereas smooth muscles squeeze consistently, the cardiac muscle is unable to sustain a contraction. Instead, the entire muscle contracts forcefully then relaxes. The cardiac muscle also looks different under the microscope than smooth muscle. Each individual muscle cell is cylindrical in shape and can have many branches. The cells join together to form long, branched tubes, with each cell separated by a disk of cell membrane called the intercalated disk.

Skeletal Muscle

The skeletal muscles are all the muscles that attach to the skeleton and help the body move. These are the 600 or so muscles that you might exercise at the gym and that you use to move around or pick up a book. They are also the muscles that get strained from exercise and that become diseased in muscular dystrophy. Most of the muscles in the body are skeletal muscles. For that reason, throughout this book the term muscles will refer specifically to skeletal muscles unless stated otherwise.

Muscles across Species

This book deals with the types of muscles in the human body, types that are shared by the other vertebrates such as birds, reptiles, amphibians, and fish. Each of these groups of animals has the three basic types of muscles that are discussed in this chapter. Other animals share some similarities with human muscular systems but also have some unique features.

Insects primarily have striated muscles like human skeletal muscles. One major difference, however, is that insect skeletal muscles attach to a hard covering on the outside of the insect's body—a shell called an exoskeleton. Other animals in the same phylum as insects include spiders, lobsters, centipedes, and millipedes, all of which have muscles connected to their exoskeleton.

With an exoskeleton, the muscles are on the inside of a limb and attach to structures on the outside, in contrast to our muscles, which are on the outside of the bone. With this arrangement, the muscle on the inside of a joint (the equivalent of the biceps) actually pulls that joint straight whereas the equivalent of the triceps bends the joint.

Shellfish such as mussels or scallops have both smooth and striated muscles. Neither insects nor shellfish need cardiac muscles because these animals have a poorly developed circulatory system. Shellfish use their smooth muscles for clamping their shell closed over long periods of time. Their striated muscles help them clamp the shell shut when they sense a disturbance such as predators. Scallops also use these striated muscles in their characteristic clamping motion as they swim. Scallops are stringier than other shellfish because they have well-developed striated muscles, which are made of stringy fibers, compared with other shellfish that primarily use their smooth muscles.

A more unusual muscular system is the hydrostatic skeleton that many soft-bodied animals such as worms use to move around. Worms have no skeleton—internal or external—for muscles to attach to. Instead, they have bands of muscles circling the body and have another set of muscles that run the length of the body. They use these muscles to push on water that fills the worm's body.

To move forward the worm would contract the circular muscles, much like squeezing a water balloon, which puts a great deal of pressure on the fluid in the worm's body. Then the worm relaxes the long muscles, and the fluid can push the worm longer in both directions. If the worm has the back part of the body anchored in some way, then this extension pushes the worm forward. A similar type of action allows worms to make complicated motions like crawling through dirt or swimming.

Skeletal muscles generally have both ends anchored to the skeleton with a thick, rope-like tissue called a **tendon**. When the muscle contracts it pulls on the tendons, which then pull two skeletal regions closer together. Usually two ends of the muscles are on either side of a joint. For example, the biceps muscle on the inside of the upper arm is attached to the shoulder at one end and to the forearm at the other end. The muscles on the inside of the joint, such as the biceps, pull two sides of the joint together, causing the joint to bend, while muscles on the outside of a joint contract to pull the joint straight.

Most muscles are organized into pairs located on either side of a joint. These two muscles work against each other and are thus called **antagonistic pairs**. One antagonistic pair might be the biceps muscle that bends the arm and the triceps muscle that straightens it back out. A similar pair is the quadriceps muscle on the front of the thigh that straightens the leg, and the hamstring on the back of the thigh that bends it. In each case, the two muscles have opposing actions and help keep the joint stable and control movement. Another type of muscle helps fine-tune direction of the movement caused by the antagonistic pair. This type of muscle works in synergy with a larger, stronger muscle and is called a **synergist**. For example, the strong bicep muscle bends the arm at the elbow, but synergist muscles control whether the hand moves directly toward the shoulder, or veers right or left.

To understand how agonists and antagonists work, imagine lifting a weight. As one lifts the weight the biceps contracts, causing the arm to bend, and the triceps relaxes to allow the bend. The triceps does stay somewhat contracted to help control how quickly the arm moves. When one lowers the weight, the triceps contracts to pull the arm straight and the biceps slowly relaxes, allowing the weight to lower.

Skeletal muscle has things in common with both smooth and cardiac muscle, but is unique in that it is the only muscle group that a person can move voluntarily. Like the cardiac muscle, each fiber can only contract once in response to a signal from a nerve. This single contraction is called a **simple twitch**. However, most movements like walking or playing basketball involve smooth, sustained motion that could hardly be achieved by simple twitches. Instead, signals from the nerves come in extremely quick pulses. Rather than resulting in a series of simple twitches, these signals cause a sustained contraction called a **tetanus contraction**. (The disease tetanus—which is also called lockjaw—causes muscles of the body to contract. See Chapter 5 for more information on tetanus.)

ANATOMY OF A SKELETAL MUSCLE

Most muscles connect to a bone at either end. The tendons that attach muscles to bones are also part of the bone's outer coating and part of the muscle's coating, making the connection extremely strong. One end of the muscle is generally considered to be the **origin**, while the other end is the **insertion**. The origin of the muscle is usually closer to the center of the body and is connected to a bone that does not move much when the muscle contracts. The insertion is usually the end of the muscle that is farther from the middle of the body and is connected to the bone that moves when the muscle contracts. For example, when the biceps muscle flexes, the forearm moves closer to the shoulder. With that in mind, the end that connects to the shoulder is the origin, while the end that connects to the forearm is the insertion.

Although most muscles are attached at both ends to a bone, a few muscles are not. These include the facial muscles that allow a person to smile or frown and the many muscles that make up the tongue. These muscles have complex origins and insertions that allow the tongue to do everything from pick food out of the teeth to control whistling and speech. Another group of muscles, such as those in the abdomen, connect to bands of tendons that cross the stomach rather than to any bone. When the muscles contract, they pull those tendons toward each other to bend the stomach.

There are also circular bands of muscles called **sphincters** that help control the flow of food through the digestive system. These sphincters don't connect to muscles, tendons, or skin, but instead form a continuous, circular band of muscles that surround the mouth, the ends of the stomach, and the anus. Puckering the lips for a kiss involves contracting a sphincter muscle. These muscles are especially good at constricting passages, such as at either end of the stomach where they prevent food from leaving until it is fully digested.

Layers of the Muscle

Although a muscle looks like a single, solid mass, it is actually made up of many smaller fibers all bundled together to form one functioning unit. Figure 1.1 shows how the many subunits of the muscle fit together. The entire muscle is wrapped in a connective tissue called the **fascia**, or the epimysium, that holds the muscle together. It's this fascia that forms into tendons and is fused to the bone to make a strong muscle-bone connection. Held within this outer covering are many smaller bundles of fibers called the **fascicles**. These bundles are also held together with connective tissue called the **perimysium**. It's these fascicles that cause the stringiness one may notice when eating meat.

Within the fascicles are the **muscle fibers** themselves. These fibers are actually very long cells—each less than the width of a hair—with many nuclei. Most cells only have one nucleus. Muscle cells are a conglomeration of many cells that fuse during development into one long fiber with several nuclei per fiber. The longest single-muscle fibers in humans tend to be about 4.7 inches (12 centimeters). The number of individual fibers in each muscle can range from 10,000 fibers to more than a million fibers, with each fiber spanning only a portion of the muscle's full length in some long muscles.

Finally, each muscle fiber contains thousands of long units called **myofibrils**. These myofibrils are what actually contract when the cell receives a contraction signal from a nerve.

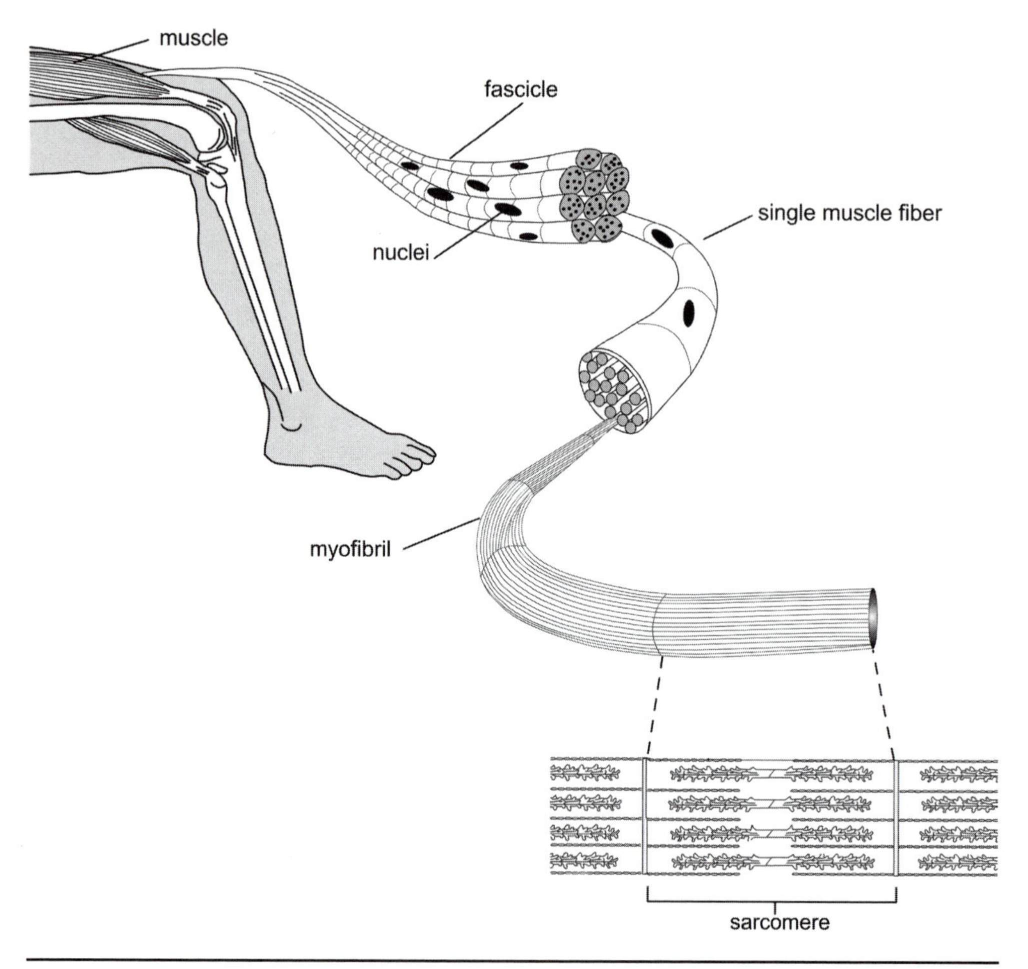

Figure 1.1. Anatomy of a muscle.

Running through the muscle between fascicles are nerves and blood vessels. The nerves relay signals that run from the brain, down the spinal column, and through nerve cells to the muscle. When a person decides to move, these nerves relay that signal to the muscle fibers that carry out that decision. The blood vessels deliver food, nutrients, and oxygen to the muscles to help the muscles contract. They also pick up carbon dioxide from the muscle and deliver it to the lungs where it is breathed out in exchange for more oxygen.

Muscle Fiber Anatomy

The individual muscle fibers are surrounded by a cell membrane called the **sarcolemma**. The sarcolemma both encircles the muscle fiber and sends hollow projections across the fiber. These channels, called **T tubules**, help the muscle transmit the signal to contract.

Just underneath the sarcolemma are granules of a type of sugar called **glycogen**. This glycogen serves as a food reservoir for the muscle to feed on when it's in heavy use. Along with the glycogen, cellular units called **mitochondria** also inhabit the space just under the sarcolemma. These mitochondria are referred to as the "powerhouse of the cell." They convert food from the bloodstream, glycogen, and other sources into energy that the muscle uses in order to contract. (Chapter 2 has more details about how the muscles convert food into energy.)

Under the mitochondria is a network of hollow tubules called the **sarcoplasmic reticulum**. The sarcoplasmic reticulum is a repository for calcium, which the muscle uses in the process of contraction. It closely follows the T tubules that cross the fiber and winds throughout the muscle fiber providing a quick source of calcium.

Ultrastructure of a Muscle Fiber

Skeletal muscle is also called striated muscle because under a microscope the myofibrils contain many tiny stripes, or striations. The stripes that can be seen under the microscope are the individual units of the muscle fiber, which are called **sarcomeres**. Cardiac muscle also has these striations, though smooth muscle does not.

Figure 1.2 shows the fine structure of the sarcomere. It is bounded on both ends by a vertical band of tissue called a **Z band**. Looking closely at the sarcomere under a microscope, there are several horizontal stripes of protein filaments. These are made up of the two primary types of protein filaments in a muscle: **actin** and **myosin**. There are roughly 1,500 myosin filaments and 3,000 actin filaments per muscle fiber.

Emanating from both Z bands are the long, thin filaments of actin. These stick out into the center of the sarcomere like posts with one end embedded in the Z band. When a muscle is at rest, these two sets of actin filaments don't overlap, leaving a gap in the center of the sarcomere.

Myosin filaments lie parallel to the actin in the center of the sarcomere and overlap with the actin on both ends. These myosin filaments are tethered in place to either Z band by a fine, thread-like filament called titin.

The bands of actin and myosin within the sarcomere account for the striations that can be seen under the microscope. The region next to the Z band that contains only actin filaments is called the **I band**. The region between actin filaments in the center of the sarcomere is called the **H zone**. The entire span of the myosin is called the **A band**.

The actin filaments are actually made up of three proteins, as shown in Figure 1.3. The globular actin proteins form into two chains that wrap around each other like two strands of twisted pearls. Winding around the actin are long, narrow filaments of a protein called **tropomyosin**. Tropomyosin fits into

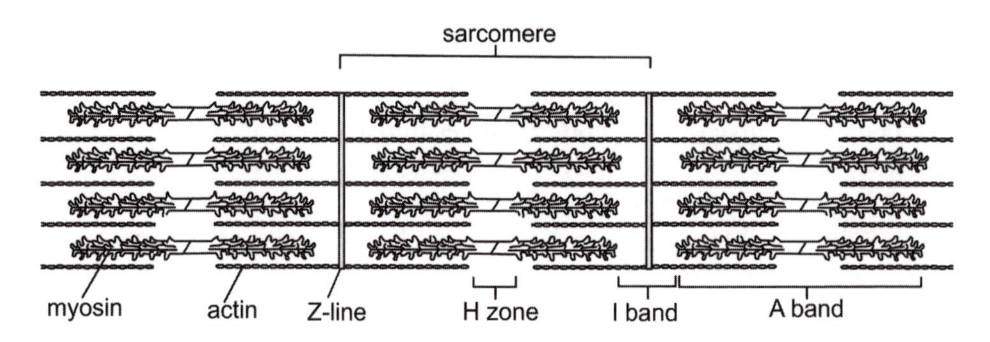

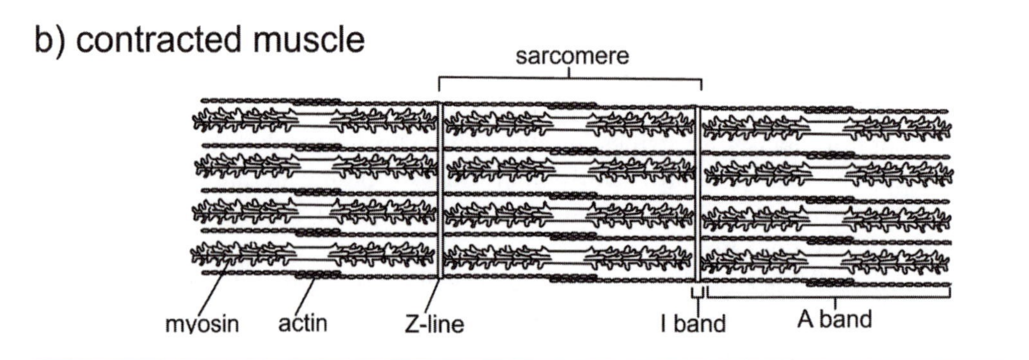

Figure 1.2. Sarcomere structure when relaxed and contracted.

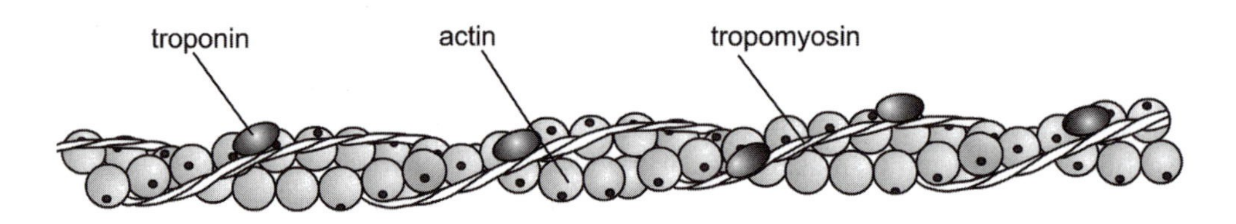

Figure 1.3. Structure of the actin filament.

the groove between the two actin strands. At intervals, a smaller protein called **troponin** dots the outside of the actin/tropomyosin complex.

Myosin is much thicker than actin and thus is sometimes referred to as the "thick filament." It looks like a chain of golf clubs twisted around each other. Each golf club represents a single myosin molecule that doubles back on itself at the tip to form the head.

ANATOMY OF A CONTRACTION

Skeletal Muscle

Each contraction results when a nerve delivers a message to the muscle. The smooth and cardiac muscles respond to nerves that are controlled by automatic reactions in the brain. Skeletal muscles move in response to nerves controlled by a person's decisions.

When a muscle gets a signal to contract, the globular myosin heads attach to the actin and pull, much like a team of people playing tug-of-war. The heads flex, release, and reattach farther up the actin filament, then flex again. This action works like a ratchet pulling the myosin along the actin filament. Because the actin filaments are attached to the Z bands at both ends of the sarcomere, this action pulls the actin filaments toward each other and pulls the Z bands closer together to shorten the sarcomere.

This mechanism is called the sliding filament model for muscle contraction. One result of the filaments sliding past each other is that the actin filaments overlap in the center of the sarcomere, causing the sarcomere to grow thicker. This overlap is what accounts for the muscle's bulge when flexed. Although each individual sarcomere only contracts a small amount, when many sarcomeres contract at once the muscle gets considerably shorter.

When a sarcomere contracts, the A band (or myosin) remains the same size in the center of the sarcomere because the myosin doesn't change length. The H band, which is the gap between the actin filaments in the center of the sarcomere, disappears as the actin filaments are pulled toward the center of the sarcomere. Likewise, the I band disappears during a contraction because the space between the Z band and the myosin filaments grows shorter. These changes within a sarcomere can be seen in Figure 1.2. Notice that as the sarcomere contracts the filaments stay the same length, although the distance between the Z bands decreases.

The strength of a contraction depends in part on how many muscle fibers (and their associated myofibrils) receive the signal to contract. When only a few fibers receive a signal, the contraction is relatively weak. When many fibers receive a signal, many more sarcomeres will contract making for a much stronger contraction.

Cardiac Muscle

Cardiac muscle has actin and myosin arranged into striated sarcomeres, much like skeletal muscle, and it has a well-developed sarcoplasmic reticulum. Unlike skeletal muscle, which contains single long fibers, cardiac muscle fibers have many branch points and appear like a complex web. When this muscle contracts, the irregular shape of the branching fibers causes the heart muscle to twist and essentially wring blood out of the heart.

Smooth Muscle

Smooth muscles lack the orderly striations of skeletal and cardiac muscles. Instead, the actin filaments are arranged in a roughly parallel way and are attached to the ends of the cell. Myosin filaments slide the actin past each other and pull the ends of the cell closer together. In smooth muscle, this action causes the entire muscle sheet to contract.

Unlike either skeletal or heart muscle, smooth muscle does not have a well-developed sarcoplasmic reticulum. When a signal from a nerve arrives at the muscle, calcium seeps in from outside the cell. This process is much slower than in the other types of muscles and doesn't allow as much calcium into the cell. Overall, the contraction takes longer to start and doesn't pull as hard as the other muscle types.

CONTRACTION STRENGTH

When a nerve sends a skeletal muscle fiber the signal to contract, that fiber contracts with an all-or-none response. It cannot only contract part way. This seems to contradict day-to-day experience, where the strength of a contraction can be adjusted for the relatively little strength required to pick up a pencil or the large amount of strength needed to pick up a large weight. If each contraction used the muscle's complete strength, then it would be nearly impossible to pick up a pencil in a controlled manner.

It turns out that the amount the muscle contracts has to do with how many fibers receive a signal. One single muscle fiber cannot produce a contraction that is strong enough to do any significant work. For this reason, most single nerves have connections with about 150 fibers, each of which contracts when the nerve sends a signal. These groups of muscle fibers are called **motor units**. In areas like the hands, where fine control is needed for activities like writing or playing a musical instrument, the motor units are composed of fewer fibers. Each nerve controls only a few fibers, providing the ability to make miniscule changes in how the muscle contracts. In muscles such as the hamstring, which provides the strength to bend the leg, the motor unit may be composed of a much larger number of fibers.

Stimulation from nerves also controls how long a contraction lasts. When a doctor taps the knee to test a person's reflexes, the leg twitches but doesn't stay contracted. That is because the muscle only receives a brief stimulation from the nerve. In order to keep the muscle contracted, the nerve sends a rapid volley of signals in quick succession. Although each signal only stimulates a single contraction, they add up so that the muscle can stay flexed to hold a paintbrush steady or hold a dance pose.

Some muscles, such as those that help a person stand upright, always have some muscle fibers contracting in order to maintain posture. In order to prevent one group of muscle fibers from getting tired, the brain sends signals to different motor units so that the groups of fibers share responsibility for holding a person upright. This same rotation takes place if one holds a heavy object for a long time. At first a few groups of muscle fibers will all get the signal to contract. But over time, if those fibers become tired the brain recruits a new group of fibers to take over responsibility for contracting. After a long time of standing upright or holding a heavy book the muscle will become tired, but no one group of muscle fibers is damaged because the motor units shared the load.

One final factor that controls the strength of a contraction is the width of a single muscle fiber. Anyone who has been to a gym knows that the more a muscle gets used, the larger it is. This change has to do, in part, with the size of the individual muscle fiber and therefore how many myofibrils are within the muscle cell. A thick muscle fiber with many myofibrils will respond with more strength in response to a signal from a nerve. In essence, that muscle cell has more myosin heads playing tug-of-war on the actin filaments, allowing that fiber to pull harder. Although the nerve will contact the same number of muscle fibers in a motor unit, the overall contraction will be stronger. Chapter 3 goes into more detail about how muscles change in response to exercise.

THE NERVE/MUSCLE CONNECTION

Skeletal Muscle

The signal to contract a muscle fiber comes from nerves. Each nerve starts at the spinal column, travels through the body, and then branches out so that the single nerve controls several muscle fibers in a motor unit. When the nerve is stimulated by the brain, an electrical signal travels the length of the nerve and reaches each of the many branched tips. At the tip, the electrical signal causes the nerve to release a chemical called **acetylcholine** into a gap between the nerve cell and the muscle fiber. The acetylcholine is also called a neurotransmitter because it transmits a signal from the nerve to the muscle.

Acetylcholine travels across the space between the nerve and muscle and binds to a protein (called a receptor) that's located on the muscle cell membrane. The bound acetylcholine triggers a series of reactions to occur that propagate the electrical signal down the muscle fiber in a wave, transmitting the signal throughout the muscle fiber and across the T tubules almost instantaneously. Where the T tubules and sarcoplasmic reticulum come in close contact, the signal from the T tubules transfers to the sarcoplasmic reticulum. In response, the sarcoplasmic reticulum releases calcium into the muscle fiber. Keep in mind that because the nerve connects to many different muscle fibers, this same reaction takes place in each fiber at exactly the same time.

Heart Muscle

The same general principles hold true between skeletal muscle and heart muscle; however, there are some notable differences. In skeletal muscle, groups of fibers within the muscle contract as they are needed—the entire muscle rarely contracts at once. In the heart, on the other hand, the entire muscle contracts with each heartbeat. For this reason, the heart muscle does not have the web of nerves branching out to connect with each individual muscle fiber.

A signal to contract reaches the heart at a specialized group of muscle cells located near the top back side of the heart. This group of cells is called the **pacemaker**. When a nerve delivers its neurotransmitter to the pacemaker region, these cells spread the signal across muscle cells in the top half of the heart. Unlike skeletal muscle fibers where the signal stays within the single fiber, heart muscle fibers are connected with an **intercalated disk** that allows a signal to move easily from one muscle fiber to the next. Because the signal can spread so quickly, all of the fibers in this top region contract at the same time. This first phase of the heartbeat pushes the blood down into the lower portion of the heart while the upper region refills with blood.

When the pacemaker receives a signal to contract, it delivers the message to all the upper muscles and also sends the signal to a group of cells in the lower portion of the heart. This message is delayed slightly to give the upper muscles time to contract. When the lower region receives the message to contract, it spreads the signal to all the muscle fibers of the lower part of the heart. This contraction finishes the heartbeat and pushes blood out to the body. The Circulatory System volume in this series has more information about how the heart relays the signal to contract.

Unlike the skeletal muscle, heart muscles can beat without a signal from the brain and can continue beating even outside of the body. Although the pacemaker cells generally relay a signal from a set of nerves, if the heart is disconnected from those nerves or if the nerves fail to fire, the pacemaker will continue sending a signal to contract at a regular rhythm. Mechanical pacemakers can back up the heart's pacemaker to ensure that the heart continues to beat at a regular rhythm, even if the heart's natural pacemaker fails to maintain the heartbeat.

SLIDING FILAMENT MODEL OF MUSCLE CONTRACTION

Skeletal and Heart Muscle

Researchers knew as early as 1883 that calcium was required in order for a muscle to contract. However, it wasn't until the 1960s that researchers understood what role calcium played when it was released from the sarcoplasmic reticulum.

It turns out that when either the skeletal or heart muscle is relaxed, the globular myosin heads cannot attach to the actin filaments. The troponin/tropomyosin complex blocks access to areas of the actin filament where the myosin head would normally bind. This physical block ensures that the muscle stays relaxed when there is no nerve signal. To prevent the muscle from contracting inappropriately, the muscle fibers actively move calcium from the cell into the sarcoplasmic reticulum, which winds throughout the cell.

When a nerve signal sweeps across the muscle fiber and through the T tubules it is translated to the sacroplasmic reticulum, which responds by releasing the stored calcium. Once in the muscle fiber, calcium binds to the troponin and causes the troponin to change shape. In this new shape, the troponin/tropomyosin complex shifts and reveals the site on the actin filament where myosin binds. Figure 1.4 shows how the troponin changes shape in response to calcium.

Although the myosin can now bind to actin, it takes energy in order for the actin and myosin to pull past each other. This energy comes from a molecule called **adenosine triphosphate (ATP)** that is normally bound to myosin when the muscle is in a relaxed state. ATP is the form of energy that is primarily produced by the mitochondria. Although ATP itself can do no work, other proteins within the cell can break the ATP into a related molecule called adenosine diphosphate (ADP) and in the process release some energy. ATP is like a match: it doesn't release any energy when it's sitting in a box, but if someone strikes the match it releases enough energy to burn fingers, melt wax, or heat a small drop of water. When the match is burned out, like ATP, it can no longer be used. Chapter 2 goes into much more detail about how muscles produce ATP and use it during a contraction.

In a relaxed muscle, myosin is bound to a molecule of ATP. When calcium enters the cell and troponin changes configuration, the myosin breaks the ATP into two units and in the process releases a small amount of energy. Myosin uses that energy to change shape and stretch out far up the actin molecule. The myosin then binds to the newly revealed binding site on actin and releases the ATP. Without ATP, the myosin relaxes back into a

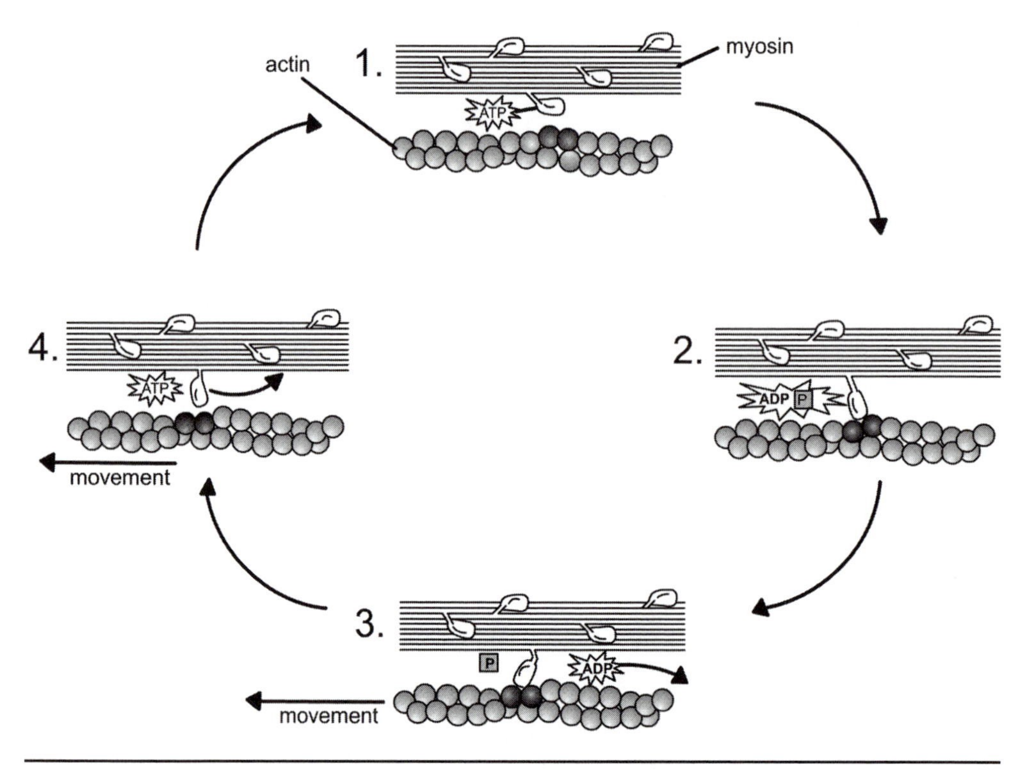

Figure 1.4. Myosin slides along the actin filament using ATP.

less energetic configuration while still bound to actin. This change in shape pulls the actin past the myosin. A new molecule of ATP then binds myosin, allowing myosin to break its bond with actin and resume its stretched-out shape, reaching farther up the actin filament.

The entire process repeats itself when myosin reattaches to the actin, relaxes, pulls the actin farther along, and binds a new molecule of ATP. With many myosin heads all attaching and flexing at different times, the process results in a slow, controlled contraction that lasts as long as ATP and calcium are present in the cell.

When a nerve stops firing, calcium stops being released from the sarcoplasmic reticulum and calcium that is in the muscle fiber is actively transported back into storage. The troponin/tropomycin/actin complex then reverts back to a shape that prohibits myosin attachment. The myosin then can't reattach, and the contraction ends. The actin filaments slide past the myosin and come back to rest with the Z bands farther apart.

One side effect of churning through ATP when a muscle contracts is that the process produces heat. People generally get hot when they exercise or use their muscles heavily—that's because each time myosin breaks ATP, a small amount of heat is released. When a person stops exercising and myosin no longer requires ATP, they eventually cool off. The body takes advantage of this phenomenon by shivering when a person gets cold. Using the muscles in this way acts as an internal heater to help warm the body. However, even with active shivering the body cannot make enough heat to warm a person when he or she is very cold. Shivering is not considered to be a useful response, because it requires quite a bit of energy to shiver with very little actual heat being produced.

Another example of the contraction process comes when an organism dies. After death, the body no longer makes ATP. Remember that ATP is needed in order for myosin to release from the actin filament after it relaxes. Once a muscle has used up its store of ATP, the myosin heads can no longer release from the actin, locking the muscles into a rigid state that is also known as rigor mortis. In this condition, all of the muscles are tightly contracted and the body is very difficult to bend.

Smooth Muscle

Smooth muscle does not contain the regular sarcomeres of skeletal and heart muscles. Instead, the actin filaments span the length of the entire muscle cell. Myosin pulls on those actin filaments to pull the cell shorter. This alternate form of contraction is also regulated in a different way. In smooth muscle, the actin filament does not have a troponin/tropomysin complex preventing myosin from binding. Instead, calcium enters the cell and activates a set of enzymes. Those enzymes add a phosphate to myosin, changing its shape and allowing it to bind actin. These intermediary steps slow the time it takes for a smooth muscle to contract. Whereas heart and skeletal muscles contract immediately when calcium enters the muscle fiber, smooth muscle cells can take as long as a second to contract after calcium levels go up in the cell.

TYPES OF MUSCLE FIBERS

Fast-Twitch and Slow-Twitch Fibers

All skeletal muscle fibers have the same basic structure, but they do vary in subtle ways that can dramatically affect the performance of the muscle. Muscles contain two general types of fibers: **slow-twitch** (type I) and **fast-twitch** (type II). As their names imply, slow-twitch fibers contract slowly whereas fast-twitch muscles contract quickly after they receive a signal from a nerve. On average, slow-twitch fibers take about one tenth of a second to reach their peak contraction while fast-twitch muscles take about half that time.

The difference between slow-twitch and fast-twitch muscle fibers lies in how quickly the myosin can cycle through ATP. The more quickly the myosin can use the ATP, the more quickly it can finish one stroke and attach to myosin for another pull.

It turns out that myosin itself is basically the same in both types of fibers. The difference between the myosins lies in their ability to break ATP—an enzyme function called an ATPase. Slow-twitch and fast-twitch fibers have myosins with different ATPase activity. As one would imagine, slow-twitch fiber myosins have an ATPase that breaks ATP slowly while fast-twitch fiber myosins have an ATPase that breaks ATP quickly.

The two types of muscle fibers also differ in how quickly the contraction begins. Fast-twitch fibers have a much more extensive sarcoplasmic reticulum network than slow-twitch fibers, allowing these fibers to receive calcium more quickly after a nerve signal than slow-twitch muscles. In the time it takes a fast-twitch muscle to flood with calcium and begin a contraction, a slow-twitch muscle is still only slowly filling with calcium, and few troponin molecules have changed shape.

Sometimes the difference between fast- and slow-twitch fibers isn't absolute. Fast-twitch fibers can be classified as type IIa or type IIb. The type IIa fibers have fast myosin ATPase and can contract faster than slow-twitch fibers, but they also have more endurance than fast-twitch fibers. The type IIb fibers are the pure fast-twitch fibers.

On average, fast-twitch fibers begin contracting five to six times faster than slow-twitch fibers. With their quick response and fast contractions, it's no surprise that sprinters have a higher percentage of fast-twitch muscle fibers than distance athletes. Chapter 2 has more information about the differences between fast- and slow-twitch fibers and how those fibers are used during exercise.

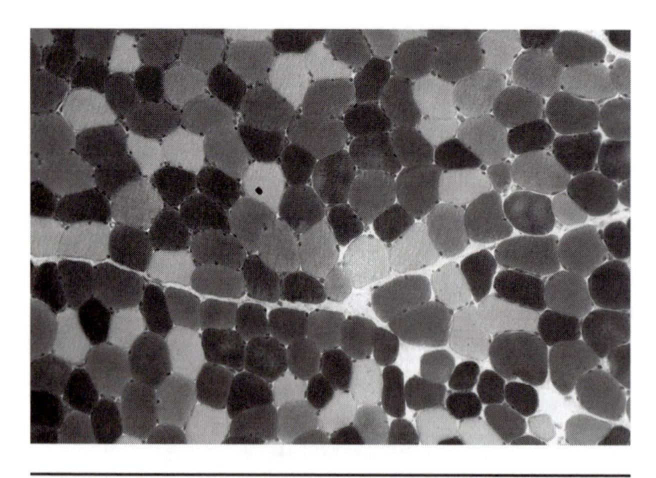

Dark stained slow-twitch fibers and light fast-twitch fibers within a muscle. © Gladden Willis/Visuals Unlimited.

Roles for Slow-Twitch and Fast-Twitch Muscles

Slow-twitch and fast-twitch fibers play distinctly different roles in a muscle. In an average muscle, slow-twitch fibers make up about half the muscle while fast-twitch fibers make up the other half. The photo shows a muscle section composed of both slow-twitch fibers (stained dark) and fast-twitch fibers. Keep in mind that people differ dramatically in their muscle composition and that fiber composition can differ even between muscles in a single person. For example, the solius muscle, which is deep inside the calf, is made up of almost exclusively slow-twitch fibers in everyone. Although the fiber composition does vary among individuals, a person who has predominantly fast- or slow-twitch muscles in the legs will have a similar composition in the arms and other muscles. This means that a person who is a particularly good sprinter when running is likely to be a sprinter rather than a distance athlete at swimming and biking as well.

When a muscle first starts being used—for example, when a person begins walking—the slow-twitch muscles are the first to be called upon to contract. When these muscles fatigue, or when the contraction needs to be more powerful, the muscle will recruit a subset of the fast-twitch type Ia fibers. Only when both of these types of fibers grow tired or when great strength is needed does the muscle call upon the fast-twitch type IIb fibers.

In order to control which type of muscle fiber contracts, each nerve connects to either all fast-twitch or all slow-twitch muscle fibers. The nerves themselves also differ. Nerves that connect to slow-twitch fibers are smaller and connect to about 10 to 180 fibers in a motor unit. A nerve for fast-twitch fibers is much larger and connects to as many as 300 to 800 fibers in a motor unit. With this arrangement, the muscle can recruit a very small number of slow-twitch muscles at a time but can signal many more fast-twitch muscle fibers with a single nerve impulse. Because of this, a single fast-twitch motor unit contraction is much stronger than the contraction of a single slow-twitch motor unit.

TYPES OF CONTRACTIONS

In general, when a muscle receives a signal to contract, the muscle grows shorter and causes a joint to either straighten or bend. Although this basic principle holds true, there are other types of muscle contraction. In most activities such as running and jumping, these different types of contraction work smoothly together to create the motion. However, it is worth discussing these actions separately to point out different mechanisms of moving the muscle.

Concentric Action

A **concentric contraction** is the motion that a person generally associates with muscle contraction. In a concentric action, the muscle contracts and pulls two bones closer together, causing the joint to open or close. Concentric action would occur if one holds a weight in the hand and bends the elbow to flex the biceps.

The myosin heads can be thought of as a team of tug-of-war players and actin as their rope. Concentric action would occur when the team pulls and the rope moves toward them.

Isometric (Static) Action

Although people usually think about contracting a muscle to move a joint, there are other types of contractions. Imagine holding a weight that is too heavy to lift. Flexing the biceps muscle, the arm will remain straight even though the muscle is trying to contract. This is an **isometric contraction**.

In an isometric contraction, the nerve sends a signal for the muscle to contract. The muscle fiber floods with calcium and the myosin pulls on the actin filaments, but the fiber does not generate enough force to bend the arm. This type of contraction also occurs when holding a weight in a stationary position. The muscle is flexing even though the weight isn't moving. Only enough muscle fibers are activated to hold the weight steady. Pushing on a wall is also an isometric action.

In the same example, if a person grows tired of holding the weight and decides to lift the weight to a new position, the muscle will recruit extra muscle fibers to help lift the load. These extra fibers then provide enough strength to lift the weight, and the isometric action becomes a concentric action. Isometric action has to do with the weight of the object and the number of muscle fibers recruited to do the work. Going back to that same team of tug-of-war players, an isometric action would occur when the players pull with all their strength but the rope doesn't budge.

The word isometric is also used to describe a type of exercise. Isometric exercises involve flexing a muscle without moving the joint. Isometric exercises were made popular in the 1950s as part of the Charles Atlas Program. Although people did become stronger using this program, their muscles were particularly stronger only when they were bent at the same angle as was used in the isometric exercises. Although isometric exercise is far from perfect, it has the advantage of requiring little equipment or room. For this reason, it is used by astronauts trying to maintain muscle strength in the weightlessness of space.

Eccentric Action

After bending the arm to lift a heavy weight (a concentric action), the arm must then straighten to let the weight down. Straightening the arm requires the biceps to maintain some contraction to control how quickly the arm straightens. This contraction that occurs as the muscle lengthens is called an **eccentric contraction**. Even though the myosin heads are pulling on the actin filaments, those actin filaments are moving farther away from each other within the sarcomere. In tug-of-war terms, an eccentric action would occur when the players pull with all their strength but the rope moves away from them and toward the other team.

Eccentric actions tend to damage the muscle fibers. Although this sounds like a bad thing, it can actually have positive results. After a session of exercise with eccentric action, the muscle fibers will become damaged and very sore. But they will adapt to just one session of eccentric exercise by increasing the number of sarcomeres per muscle fiber (essentially making each sarcomere shorter). With more sarcomeres per fiber, each sarcomere has to stretch less for the eccentric action, so the muscle sustains less damage. The other advantage to this is that the muscle will become quicker to respond to a signal because there are many more sarcomeres doing the same work. This change lasts about ten weeks—after that time, the muscle loses the adaptation and will once again be damaged by eccentric action.

Most people have experienced pain associated with eccentric action when they run or walk downhill. When a person walks downhill, the thigh muscle contracts as the leg extends in front. Transferring weight onto the other leg, the knee lowers slightly to absorb the shock and in the process stretches the thigh muscle. Stretching the thigh muscle while the muscle contracts is a form of eccentric action that damages the muscle and can lead to pain and weakness over the next two to three days. Walking or running downhill in the next few weeks will cause much less pain in the following days because the muscles have adjusted to the eccentric action. Chapters 3 and 8 go into more detail about eccentric action during exercise and how to treat muscles that are injured by eccentric action.

2

Energy Use by Muscles

Muscles require energy in the form of adenosine triphosphate (ATP) in order to contract. This ATP is used as an energy source for all the reactions that take place in the body—it is required to conduct signals along the nerves, to translate genes into proteins, to move molecules into and out of a cell, and to contract muscles, among other things.

Each cell is self-sufficient when it comes to making ATP. A person cannot eat ATP and have the molecule transported to cells, nor can energy-starved cells take ATP from the bloodstream. Instead, each cell generates its own ATP from sugar, fat, or protein that comes either from the bloodstream or from the cell's internal stores. For this reason, cells that require large amounts of energy, such as muscle cells, have to be extremely efficient at making ATP and at storing the starting materials.

The Respiration volume of this series provides detailed information about the process of how cells convert food into ATP. This chapter focuses on those details of respiration that are important in order to understand how muscles get the energy to contract and what happens as muscle cells use up their energy stores.

GETTING ENERGY FROM ATP

ATP belongs to a class of molecules called nucleotides—the same molecules that make up DNA. Like other nucleotides, ATP is made up of three parts: a sugar called ribose, a double ring of carbon atoms called adenine, and three phosphates attached like a tail. These phosphates are linked by high-energy bonds. When a cell needs energy, it breaks off one of the phos-

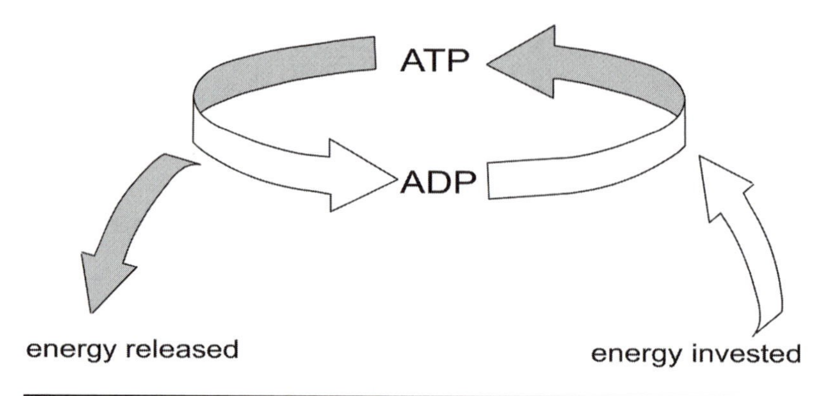

Figure 2.1. Conversion between ATP and ADP.

phates from the ATP to form adenosine diphosphate (ADP) and one free phosphate. Removing that phosphate releases enough energy to drive a cellular reaction such as flexing the myosin head. Because the remaining ADP is not a source of energy, a cell needs to constantly convert ADP back into ATP by adding a phosphate back to the tail—particularly if that cell requires a large amount of ATP to do its job, such as a muscle cell. Figure 2.1 shows the cycle of using energy to create ATP, then releasing energy when ATP is converted into ADP.

It may seem surprising that breaking off a piece of a molecule can result in usable energy. However, the process is similar to how cars get energy out of burning gasoline or how fires produce heat. In each case, breaking the bonds between two atoms releases energy. In a car engine, breaking the bonds of atoms in gasoline makes energy to drive the pistons that turns the motor. In a fire, that energy takes the form of heat; and in ATP, the energy is used for cellular reactions. In each of these examples, oxygen is needed in order to break the bonds in the molecule and carbon dioxide is released.

CREATING ATP

Generating ATP is a several-step process. The first phase takes place in the main cavity of the cell, whereas the second—and most ATP-producing—step takes place in a cellular component called the mitochondria.

A mitochondrion is a small, bacteria-shaped component of the cell. Each cell has from one to many mitochondria, depending on how much energy it needs in order to survive. It is no accident that the mitochondria resemble bacteria—researchers think that hundreds of millions of years ago, free-living bacteria entered larger cells. These bacteria received food from the cell, and in return they provided ATP. Over time, the bacteria lost their ability to exist without their host cell and became the mitochondria that cells have today. Mitochondria still have some remnants of their free-living bacterial ancestors, such as some DNA and genes that make bacterial proteins. The process mitochondria use to make ATP is similar to how free-living bacteria still function.

Glycolysis

The first step of producing ATP—called **glycolysis**—takes place outside of the mitochondria in the main compartment of the cell. This process requires a sugar called glucose as a starting material. The glucose molecule is a chain of six carbon atoms with some hydrogen and oxygen ($C_6H_{12}O_6$). It can either come from the bloodstream, where glucose constantly circulates to feed cells, or from the muscle cell's internal stores.

During the glycolysis process, a series of reactions splits glucose into two halves, each of which is called **pyruvate** (in fact, the name glycolysis means "sugar splitting"). For each sugar that enters glycolysis, the process makes two molecules of pyruvate and two molecules of ATP. Although this is a net gain in ATP, two ATP molecules are not enough to fuel extensive exercise.

Citric Acid Cycle

The next step in the ATP-producing pathway takes place within the mitochondria. Because of their critical importance in creating energy for the muscle, mitochondria are located underneath the sarcoplasmic reticulum, overlaying the muscle fibers themselves. In this position the mitochondria are poised to receive fat or sugar from the blood and convert that into ATP. In most cells, mitochondria exist as individual units that dot the interior of the cell. They are so important to a muscle cell, however, that they fuse to form a single large entity that spans the entire muscle cell.

The process that takes place within the mitochondria is called the **citric acid cycle**—also called the Krebs cycle after Sir Hans Krebs (1900–1981) who first discovered the cycle in the 1930s. The cycle is also sometimes called the tricarboxylic acid (TCA) cycle because of the three-carbon molecule that continually moves through the cycle. The citric acid cycle begins when pyruvate from glycolysis loses one carbon atom, to become a two-carbon molecule called **acetyl coenzyme A (acetyl CoA)**. This acetyl CoA then enters the mitochondria, where it goes through eight individual steps that produce carbon dioxide that gets breathed out, one more molecule of ATP, and two high-energy molecules called NADH and $FADH_2$. These are the molecules that have the most potential to generate ATP.

Although each acetyl CoA that enters the citric acid cycle produces one ATP, every glucose produces two molecules of acetyl CoA. This means that the citric acid cycle produces two molecules of ATP per glucose, bringing the total up to four molecules of ATP for each glucose that goes through glycolysis and then into the citric acid cycle. Again, four molecules of ATP is better than no ATP at all, but it isn't enough to sustain long-term exercise such as jogging or swimming.

Sugar is the most common fuel for making ATP, but cells can also get energy from fat in the form of free fatty acid in the blood. A fat molecule is a very long chain of carbon atoms. One common fat that is found in the cell is called palmitate, which is a chain of sixteen carbons. When a fat molecule enters the cell, it is broken down in two-carbon units into acetyl CoA—the same molecule that sugar is converted into after glycolysis. This acetyl CoA then goes through the citric acid cycle and produces carbon dioxide, one ATP, and NADH and $FADH_2$.

With sugar, scientists can calculate how many ATPs come from each molecule. That's because each sugar has six carbons and sends two acetyl CoA molecules through the pathway. Fat molecules can be different lengths, so it is harder to calculate how many acetyl CoA molecules will be made from one fat molecule and therefore calculate how much ATP will be produced. Regardless, a fat molecule will produce the same amount of ATP, carbon dioxide, NADH, and $FADH_2$ per acetyl CoA as each acetyl CoA from sugar—it is just a matter of how many acetyl CoA molecules can be generated from each fat molecule.

The difference in how fat and sugar molecules enter the citric acid cycle helps explain why a gram of fat has so many more calories than a gram of sugar. A **calorie** is essentially a measure of how much energy that food contains. From each six-carbon sugar, only four carbons go through the Krebs cycle (two molecules of two-carbon acetyl CoA). The remaining two carbons are breathed out as carbon dioxide. With fat, the entire molecule is broken down in two carbon chunks, so the entire molecule is used to create ATP. Because the entire weight of the fat molecule goes to making energy, it can provide more energy per weight than sugar. This translates into more calories per gram on a food label. The benefits of burning fat over burning sugar for energy is not lost on muscle cells. As muscles become highly trained, they also become better at using fat for energy.

Electron Transport

The final phase of ATP production involves the NADH and $FADH_2$ that were made during both the citric acid cycle and glycolysis. These molecules both carry extra electrons that are in a high-energy state. They transfer their electrons to a series of proteins that are lodged in the membrane of the mitochondria. Together, these proteins are called the **electron transport chain**. They form a continuous path for electrons, picking up the electrons from NADH and $FADH_2$, and transferring them down the chain. This step leaves the NADH and $FADH_2$ short on electrons. The depleted molecules go back to the citric acid cycle where they pick up new high-energy electrons that can once again be donated to the electron transport chain.

In the final step of the electron transport chain, the electrons combine with oxygen in the mitochondria to form water. Oxygen is essential to this step. Without oxygen, electrons pile up in the electron transport chain and transport stops. When this happens the citric acid cycle also grinds to a halt, pyruvate stops being imported into the mitochondria, and all the ATP must be made by glycolysis.

The point of moving electrons along the mitochondrial membrane is not simply to convert an oxygen into water—at each step of the chain, the same reaction that transfers the electron to the next protein also transports hydrogen ions from inside the mitochondrial membrane to the outside of the membrane. These hydrogens build up on the outside of the membrane like water behind a dam. When the mitochondria releases the hydrogen ions back inside (through a molecule called the ATP synthetase), those hydrogens do the equivalent of the spinning wheels inside a dam. In dams, the spinning wheels make electricity that powers homes. In a cell, unleashing the dam produces about thirty-two molecules of ATP per sugar molecule. The process of producing ATP through the electron transport chain is called **oxidative phosphorylation**.

The entire process of breaking down sugar in glycolysis, sending pyruvate into the mitochondria, and transporting electrons down the electron transport chain is shown in Figure 2.2. Figure 2.3 shows an overview of where the processes take place in the cell.

The Importance of Oxygen

Only four of thirty-six total molecules of ATP come from glycolysis and the citric acid cycle, while the remaining thirty-two molecules come from oxidative phosphorylation. Oxidative phosphorylation is a rich source of ATP, as long as the mitochondria receives a steady supply of oxygen.

The oxygen used in the electron transport chain accounts for why a person's breathing and pulse increase when they exercise. If a person can't deliver enough oxygen to the muscles, then the electron transport chain shuts down and the muscle only receives the paltry amount of ATP made by glycolysis. For low-level exercise such as gardening, a person will breathe slightly harder than usual in order to provide the muscle cells with enough oxygen for their slightly increased ATP demands. A person in a swimming race, on the other hand, must breathe very hard and pump new oxygen-containing blood to the muscles very quickly to keep up with the muscle's high demand for ATP.

Exercise that takes place through ATP made by oxidative phosphorylation is called **aerobic exercise** (aerobic means "with oxygen"). Exercise that

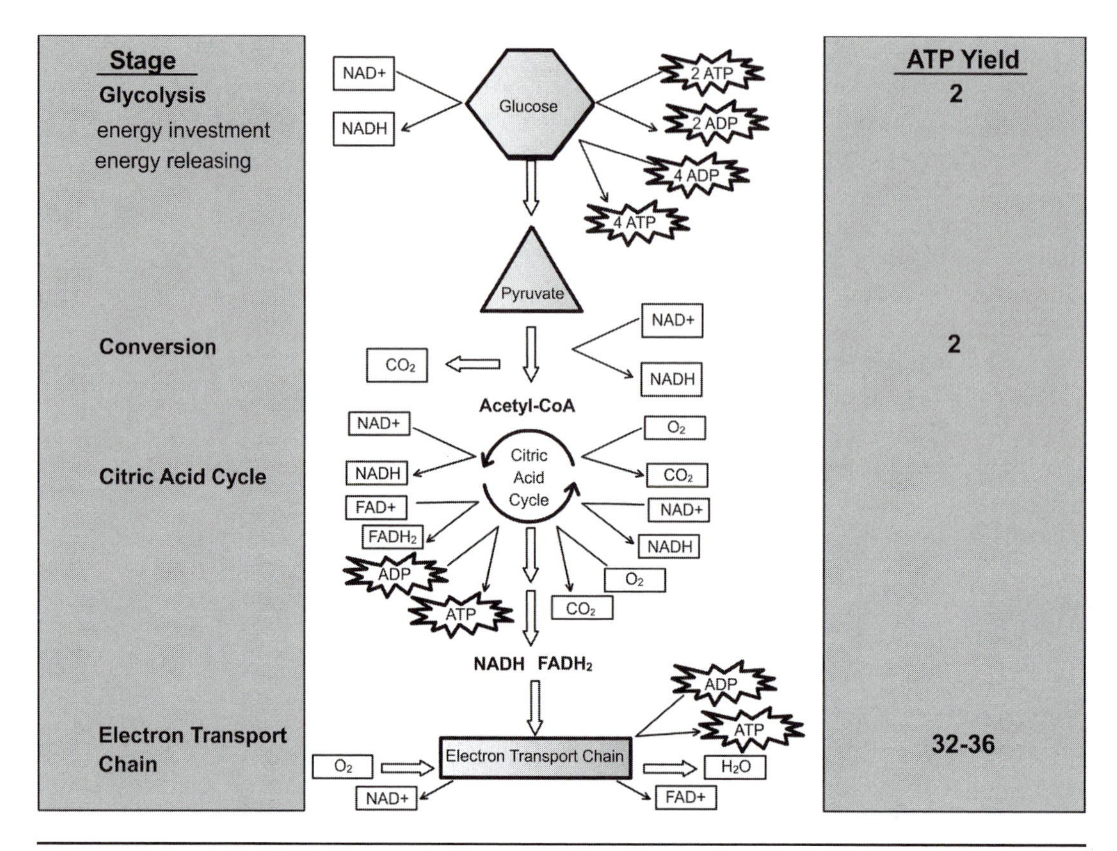

Figure 2.2. Overview of energy production from glucose.

takes place when oxygen is limited and all ATP comes from glycolysis is called **anaerobic exercise** (anaerobic means "without oxygen").

Once a person is done exercising, his or her breathing and heart rate stay elevated for several minutes even though the muscles are no longer using ATP. This happens because the muscle cells need to replenish their supply of stored-up ATP. A person will continue breathing faster than usual until the muscle has enough ATP stored up to act as a buffer against future bursts of exercise.

SOURCES OF ENERGY

Cells in the body can convert sugars, fat, or protein from the bloodstream into ATP. Although the bloodstream provides a continuous source of fuel, muscles also store their own fuel in the form of molecules called **creatine phosphate** and glycogen.

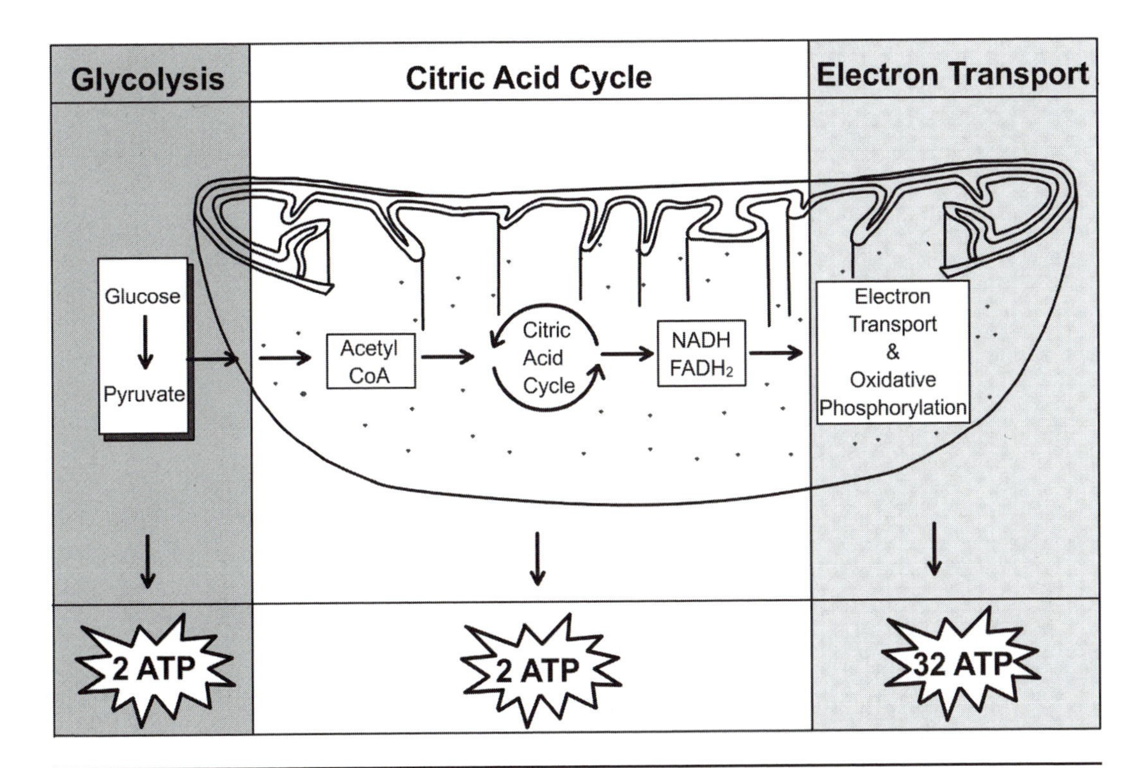

Figure 2.3. ATP produced from glycolysis, the citric acid cycle, and electron transport.

Creatine Phosphate

Researchers knew from early experiments that a backup energy source must exist in muscles. In these experiments, Eggleton and Eggleton removed a piece of muscle but kept the nerve intact so that it could still stimulate the muscle to contract. When they gave that muscle chemicals to prevent either glycolysis or oxidative phosphorylation, then stimulated the nerve, the muscle could still contract. This result told the researchers that the muscle must have some way to regenerate ATP that did not include either glycolysis or oxidative phosphorylation.

In 1927 the researchers identified this short-term energy store is a molecule called creatine phosphate. Creatine phosphate serves as a reservoir for phosphates that can be added to ADP to regenerate ATP. When the cell is at rest and making excess ATP, then the ATP will transfer one phosphate to creatine to make creatine phosphate, converting the ATP to ADP. (The ADP receives a new phosphate through glycolysis or oxidative phosphorylation and is converted back into ATP.)

When a person begins exercising—for example, by suddenly standing up and beginning to walk briskly—the muscles don't have enough time to ramp up ATP production. These muscles will use up all the available stored ATP in about two seconds. That is less time than is needed for the muscle to begin producing energy through glycolysis and oxidative phosphorylation. During the lag time between when muscles use up stored ATP and when new ATP can be created, the creatine phosphate donates phosphate back to ADP to regenerate ATP and maintain energy for a contraction. By the time the cell depletes its store of creatine phosphate, the mitochondria have begun churning out a fresh supply of ATP.

When the exercise ends, the muscles refill the bank of creatine phosphate to be used at a later time. This refueling can happen on the order of a few minutes, which is why a sprinter or weightlifter can do several repetitions of creatine phosphate–depleting exercise with only a few minutes to recover.

Glycogen

After eating a meal, digestive enzymes in the stomach and small intestine break down food into sugars, protein, and other nutrients that are absorbed into the blood from the small intestine. For about an hour after eating a meal, the blood has a lot of sugar in the form of glucose, which the cells of the body take up and use to produce ATP or store for later use. However, the body eventually uses up all that sugar and must turn to another source of energy.

Cells of the liver and muscle save sugar to use when supplies are limited between meals. They both take up excess sugar when it is plentiful and convert it into chains of six-glucose molecules called glycogen. In the muscle, the sarcoplasmic reticulum serves as a reservoir for this glycogen. When the muscles need energy, they can convert the glycogen back into sugar and feed that sugar into the energy-producing pathway to produce ATP. The liver releases its glycogen as sugar into the blood to keep a steady supply of sugar available to other cells in the body, including the brain.

Where creatine phosphate provides a short burst of energy for cells, it is quickly depleted. After the creatine phosphate is used up, the muscles turn to glycogen stores to create new ATP. In most cases, this stored glycogen will provide enough energy for any activity the muscle needs to accomplish, such as walking, running, or housecleaning.

Fat

When blood sugar levels are high after a meal, liver and muscles cells aren't the only ones planning ahead. Fat cells take up sugar and convert it into a form of fat called a triglyceride, which the cell stores as a single, large droplet for later use. This fat is a reserve against the future when the body needs fuel and all the creatine phosphate and glycogen are used up. The fat cells then release the stored fat into the blood in a form called a fatty acid, which the mitochondria can convert into ATP. In addition to fatty acids in the blood, muscles cells store tiny droplets of fat that they can use during exercise.

Although the muscles use creatine phosphate, glycogen, and fat more or less in sequence, it is important to note that the body almost always has some sugar and some fatty acid present in the blood at any time. Muscles will also use these fuel sources in different quantities under some conditions. As an example, highly trained endurance athletes rely more heavily on fat for energy—an adaptation that makes the muscle far more efficient at getting the most energy out of the lowest-weight fuel source. Moreover, the brain can only survive on sugar, so for normal brain function there must always be some sugar in the blood.

Protein

Most food is made up of sugars (carbohydrates), fat, and protein. Although the body stores sugar and fat for later use, it does not store excess protein. Instead, the protein subunits (called amino acids) are taken up from the bloodstream and used to build new proteins in cells. Some of that protein can also be taken up by liver cells and converted into sugar that then fuels ATP production in cells or is converted into fat by fat cells.

Protein in the bloodstream is very rarely used as an energy source. During starvation, the body may break down existing proteins and use that to fuel the production of ATP. But in a normal situation, protein contributes only about 5 percent of the fuel to make ATP. Some protein subunits can be broken down into acetyl CoA and fed directly into the citric acid cycle. Other subunits feed in partway through the cycle and only produce some of the ATP, NADH, and $FADH_2$ that are normally made in the process.

REGULATING THE FUEL SOURCE

Which source of energy the muscle cell uses depends on both what type of work the muscle is performing and what energy source is available. For short, intense work, the muscle will rely on creatine phosphate and muscle glycogen. But for longer work, it may rely on a combination of muscle glycogen, sugar and fatty acids from the blood, and even protein. A number of factors determine what energy source a muscle uses, including diet, caffeine, hormones, and the environment.

Sugar Availability

After exercise that uses up all of the stored glycogen—such as running a ten-kilometer race—it takes about twenty-four hours before the muscles can refill their glycogen supplies. If the person uses the muscles before they have replenished the glycogen supplies, the muscle will have to rely more heavily

Eating Carbohydrates Improves Endurance Performance

Athletes who are engaging in long-distance events rely largely on muscle glycogen for energy—in addition to fats and sugar from the blood. It's this glycogen that athletes are trying to increase when they **carbo-load** (eat foods that are high in carbohydrates such as potatoes or pasta) before a race. Muscles that are highly trained will take that burst of carbohydrates and store it as excess glycogen.

When athletes train for long-distance events such as a marathon, their blood supply, mitochondria numbers, and glycogen stores all increase to help those slow-twitch muscles continue working effectively throughout the event. Chapter 3 will go into more detail about muscular changes that occur through training.

on fat for energy. People who eat a high-fat diet will have more fat than sugar available in the blood and will likely use fat more extensively to generate ATP. (See also "Eating Carbohydrates Improves Endurance Performance.")

Caffeine

One of caffeine's many side effects is that it causes fat cells to release fatty acids into the blood. With higher blood concentrations of fatty acids, the muscles will rely more on fat and less on stored glycogen for energy. The effect is particularly relevant in the first twenty or so minutes of exercise, when muscles usually use very little fat. By drinking caffeine before an event, athletes can increase their endurance by using fat for energy early on and sparing the muscle glycogen for later.

Hormones

A combination of hormones in the body regulates the sugar and fat supplies to ensure that the muscles have a steady supply of fuel. Hormones such as adrenaline—which the body secretes under times of stress and during exercise—cause the body to release more fatty acids and therefore to use fatty acids for energy in the muscles. Adrenaline also causes the liver to break stored glycogen into glucose and release that into the blood.

Two other key hormones regulate how sugar is used by the body. The pancreas releases **insulin** after a meal when sugar levels are high. During this time, muscles take up sugar to store as glycogen and are more likely to use sugar to make energy. When sugar levels fall, the pancreas stops making insulin and instead makes **glucagon**. Glucagon causes the fat cells to release fatty acids and the liver cells to release glucose. During exercise the body

Exercise Improves Type II Diabetes

To take advantage of the glucose in the bloodstream, muscles that are exercised regularly become better able to take up glucose from the blood. This effect is also why people with Type II diabetes are encouraged to exercise regularly. In Type II diabetes, people become increasingly resistant to insulin, which is the hormone that helps cells of the body take up sugar. With the cells not able to take up sugar, blood sugar levels increase and cause damage to the eyes, kidneys, nerves, and blood vessels. Through regular exercise, muscles in people with Type II diabetes take up more sugar from the blood, lowering blood sugar levels and helping to prevent long-term damage to the organs. Exercise can also cause a person to lose weight, which also helps control Type II diabetes.

produces additional glucagon, which further increases the amounts of fat and sugar that are available to the muscles. For more information about how insulin and glucagon regulate blood sugar levels, see the Endocrine System volume of this series. (See also "Exercise Improves Type II Diabetes.")

With so many hormones all telling the liver to release glucose during exercise, blood glucose levels can be as much as 50 percent higher after a short period of exercise. This phenomenon is in part responsible for why people often don't feel hungry immediately after exercising. For most athletic events, the liver releases glucose at about the same rate that the muscles take it up. But for very long events, the liver may run out of glycogen and not be able to keep pace with the muscle's glucose demands. At this time, the muscles rely entirely on fat or on food such as energy bars or gels.

Environment

Heat and altitude can both cause the body to use carbohydrates for energy and to rely more on glycolysis for ATP. At higher altitudes, the air contains less oxygen and therefore less oxygen is present in a person's bloodstream. With limited oxygen, the muscles cannot generate enough ATP through oxidative phosphorylation and must instead use glycolysis for the remaining ATP. This effect can leave a person winded and unable to keep exercising as long as would be possible at lower altitudes.

A similar phenomenon happens in the heat. The body diverts blood to the skin where it can radiate heat, but this takes blood volume away from the muscles. With less oxygen, fatty acid, and sugar being delivered, the muscles must use stored glycogen to make energy through glycolysis. As with exercise at a high altitude, a person exercising in the heat will have less endurance than on a cooler day when the muscles generate ATP though oxidative phosphorylation.

Energy Use during Sprint and Endurance Exercise

During long-distance events such as a ten-kilometer running race, the muscles receive enough oxygen from the blood and generate ATP through oxidative phosphorylation. In one study, researchers found that marathon runners rely 99 percent on aerobic forms of generating ATP, whereas in a 100-meter sprint 90 percent of the energy comes from anaerobic processes. An 800-meter sprint lies between these two extremes, with 60 percent of energy coming from anaerobic means.

GENERATING ENERGY IN DIFFERENT MUSCLE TYPES

There are two general types of muscle fibers—slow twitch and fast twitch. The slow-twitch fibers are specialized for endurance use, while fast-twitch muscles excel at short, intense bursts of strength. With their respective specialization, the two muscle types differ in how they generate energy and how effectively they use oxygen. (See "Energy Use during Sprint and Endurance Exercise.")

Energy in Slow-Twitch Muscles

Muscles use slow-twitch fibers for sustained exercise such as walking or dancing or even running a marathon. These activities rarely involve large amounts of strength but do require the muscle to contract repeatedly over a long time. In order to generate enough energy to contract over such a long time period, slow-twitch muscles must be very effective at delivering oxygen so the muscles can maintain oxidative phosphorylation. Any type of endurance exercise that requires large amounts of oxygen is called aerobic exercise.

Blood vessels threading throughout the slow-twitch muscle fibers ensure that the muscle gets enough oxygen to make ATP effectively. These blood vessels bring oxygen from the lungs and return carbon dioxide created through the citric acid cycle. Blood vessels also take up water that is made in the final step of the electron transport chain. In addition to having blood vessels bring in plenty of oxygen, slow-twitch muscles contain a protein called **myoglobin** that is extremely effective at removing oxygen from the blood. This myoglobin gives the fiber a red color, which was originally used by physiologists to distinguish between fast- and slow-twitch muscle fibers.

To make the best use of all the oxygen coming in from the blood vessels, slow-twitch muscles contain many more mitochondria than fast-twitch muscles. In fact, in some muscle fibers these mitochondria make up 20 percent of the muscle fiber volume.

Energy in Fast-Twitch Muscles

Fast-twitch muscles are used for bursts of speed—exercise that ends before the heart can begin beating faster to deliver more oxygen to the muscles. In order to generate enough ATP for the contraction without being able to rely on the electron transport chain, fast-twitch muscle fibers are adapted to be extremely efficient at glycolysis. They have fewer blood vessels and mitochondria than slow-twitch muscles and store more creatine phosphate in order to replenish the ATP supplies. Fast-twitch muscles also contain more of the molecules that carry out the process of glycolysis and store far more glycogen than slow-twitch muscles to ensure a continuous supply of sugar to feed into glycolysis.

As discussed in Chapter 1, there are two types of fast-twitch muscles—type IIa and type IIb. Type IIb fibers are the ones that are most specialized for fast, sprint-type work. They have the fewest mitochondria, least myoglobin, and most glycolysis enzymes of any muscle fiber.

The type IIa fibers are somewhere between the slow-twitch fibers and type IIb fibers. Their myosin heads pull and release the actin filaments quickly, like the IIb fibers, and they are used for sprinting rather than endurance exercise. They have more mitochondria than IIb fibers, however, and are also darker in color because they contain some myoglobin. These fibers often serve as a backup when the slow-twitch muscles become fatigued. Table 2.1 shows the differences between the different muscle types. (See "Light and Dark Meat Represent Fast- and Slow-Twitch Muscles.")

Light and Dark Meat Represent Fast- and Slow-Twitch Muscles

The differences between fast- and slow-twitch muscles are noticeable by the naked eye as light and dark meat in chicken or turkey. The dark meat on the thighs is juicier because of all the fatty membranes around the blood vessels and mitochondria. The difference in color comes from a protein called myoglobin that helps the muscle extract oxygen from the blood.

In poultry, the thighs are used all day long for walking and standing, so these muscles have a large number of slow-twitch fibers and are dark in color. The lighter muscle that makes up the breast meat is used for short bursts of flying—exercise that takes a lot of strength but doesn't last long. This muscle is dryer because it has less fatty cell membranes surrounding the blood vessels and mitochondria and is lighter in color because it has less myoglobin.

TABLE 2.1. Differences Between Muscle Fiber Types

	Slow-Twitch	Fast-Twitch IIa	Fast-Twitch IIb
Contraction speed	Slow	Fast	High
Number of mitochondria	High	Medium	Low
Glycolysis enzymes	Low	High	High
Citric acid cycle enzymes	High	Medium	Low
Creative phosphate levels	Low	High	High
Motor unit strength	Low	High	High
Endurance	High	Moderate	Low

FATIGUE

Glycolysis, the citric acid cycle, and the electron transport chain all provide energy to the cell, and each muscle is specialized to use these processes effectively for a specific type of exercise. But when the creatine phosphate and glycogen run out, muscles struggle to provide enough energy to maintain the exertion. This state of not being able to provide enough energy is called fatigue. Slow-twitch and fast-twitch muscles rely on different processes to generate energy and therefore handle fatigue differently.

Aerobic

In longer-distance events, an athlete starts out at a slow pace and runs through only a small portion of the creatine phosphate before the heart rate and breathing increase and the muscles receive enough oxygen. These athletes transition into generating ATP through oxidative phosphorylation before the creatine phosphate runs out.

When physiologists measure muscle glycogen levels while athletes perform endurance exercise such as riding a stationary bike, those athletes report being fatigued around the same time that their glycogen levels fall to nearly zero. Marathon runners often report hitting a wall near the end of the run when they suddenly feel extremely fatigued. This wall corresponds with when muscle glycogen is used up. It is the state of fatigue that athletes try to avoid by eating a lot of carbohydrates and building up glycogen supplies the night before a race. (See "Glycogen Loss Is Visible under a Microscope.")

When the slow-twitch muscles no longer have stored glycogen, they can't generate enough ATP and can no longer contract properly. Remember that muscle fibers are combined into groups of motor units. As fibers in the first motor unit fatigue, the body will recruit a new group of slow-twitch fibers.

Glycogen Loss Is Visible under a Microscope

Exercise physiologists can look at glycogen stores in the muscle to tell which muscle fibers have been used extensively during a particular type of exercise. First they take a sample of the muscle, called a **biopsy**, and use a stain that distinguishes fast-twitch from slow-twitch muscles. Another stain indicates which fibers contain glycogen. As expected, during long-distance exercise the slow-twitch muscles are depleted in glycogen while fast-twitch muscles still contain glycogen reserves. After more intense forms of exercise, fast-twitch fibers burn through their glycogen at a much faster rate than the slow-twitch fibers.

If the exercise goes on long enough, all of the slow-twitch fibers will end up fatigued and will be unable to contract. The muscle compensates for this loss in strength by recruiting fast-twitch IIa fibers, then recruiting fast-twitch IIb fibers. Because fast-twitch muscles aren't efficient at aerobic activity, this transition can leave the athlete feeling fatigued and heavy.

The slow-twitch muscle fibers don't rely on stored glycogen alone—sugar in the blood feeds into the glycolysis pathway and fat enters directly into the citric acid cycle. During exercise, the muscles rely on both of these fuel sources to generate ATP. The sugar comes primarily from the liver, which also stores glycogen. When blood sugar levels fall, the liver breaks glycogen down into usable sugar, and this sugar enters the bloodstream where it eventually reaches the muscles. Athletes who run low on muscle glycogen rely extensively on energy sources in the bloodstream.

For long-distance swimming, biking, or running events, the athletes cannot rely exclusively on stored glycogen, and liver glycogen may also run low. In fact, for events such as a marathon, the body generally does not carry enough total sugar in the muscle and liver to generate enough ATP for the entire race. This is why many athletes eat energy bars or gels during a workout. These food sources provide a quick source of sugar to replace the glycogen the athlete has used up.

Anaerobic

As long as the muscle gets enough oxygen, the electron transport chain should continually provide enough ATP for the muscle to continue functioning. But during some very intense forms of exercise, such as weightlifting or sprinting, the muscle requires an immediate burst of ATP. In these types of events, the muscles don't get enough oxygen to sustain oxidative phosphorylation and must rely on ATP from the pool of creatine phosphate and from glycolysis. Muscles can only rely on ATP from glycolysis for about twenty seconds to five minutes, which is why fast-twitch muscles can only sustain exercise for a short period of time.

When exercise physiologists look at muscles in people who are doing sprint events, those athletes report being fatigued around the same time that their creatine phosphate levels run low. At this point, the muscles have no steady source of ATP except through the relatively inefficient glycolysis, which is relatively inefficient and can't provide enough ATP to sustain an intense workout.

There are two downsides to relying on glycolysis for ATP. The first is that the process doesn't produce very much ATP. In addition, the end product of glycolysis—pyruvate—gets converted into a related molecule called **lactic acid** that can cause the muscle to become acidic.

The purpose of converting pyruvate into lactic acid is to regenerate one of the cofactors that gets used up during the process of glycolysis. This molecule, called NAD^+, is normally regenerated in the electron transport chain. When the cell lacks oxygen and isn't using the electron transport chain, NAD^+ can be in short supply. The process of converting pyruvate to lactic acid regenerates NAD^+ so that glycolysis can continue. Figure 2.4 shows the

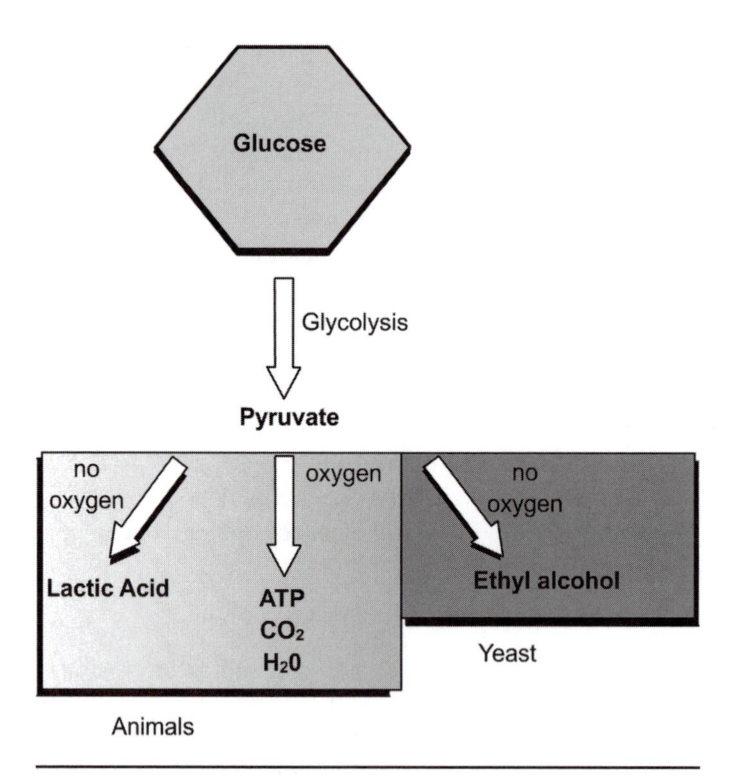

Figure 2.4. Pathways for pyruvate break-down in animals and yeast.

Yeast Glycolysis Produces Beer and Bread

In most animal muscles, pyruvate gets converted into lactic acid in order to regenerate NAD^+ when there isn't enough oxygen for the citric acid cycle. However, yeast uses an alternate method to regenerate NAD^+—one that has benefited humans for many millennia.

Yeast have no mitochondria and use glycolysis exclusively to generate ATP. When pyruvate builds up at the end of the process and NAD^+ runs low, the yeast regenerate that NAD^+ by converting pyruvate into ethyl alcohol rather than into lactic acid. This ethyl alcohol is the alcohol found in alcoholic beverages.

The process of making alcohol such as beer or wine involves feeding a form of sugar to yeast, then letting the yeast convert those sugars into alcohol. The starting product determines the type of beverage that results—malt produces beer, grapes produce wine, potatoes produce vodka, rice produces sake, and rye produces bourbon.

Yeast is also used to help bread or pastries rise. In this process, the yeast breaks down sugar in the dough to produce its ATP. The yeast also converts the pyruvate into alcohol, which evaporates during baking, and the yeast produces carbon dioxide during glycolysis. This carbon dioxide bubbles up through the dough and causes the dough to rise.

possible pathways pyruvate can take following glycolysis in humans and in yeast. (See "Yeast Glycolysis Produces Beer and Bread.")

During anaerobic exercise, lactic acid builds up in the muscles, enters the bloodstream, and is transported to the liver. Once in the liver, two molecules of lactic acid are recombined to form glucose in a process called gluconeogenesis. This new molecule of glucose can then be stored in the liver as glycogen or goes out into the bloodstream where it can be used by any cell in the body—including the muscle—for energy. The process of glucose being converted into lactic acid, then transported to the liver where it is recombined to form glucose again, is also called the Cori cycle, named after Carl Ferdinand Cori (1896–1984) and his wife Gerty Theresa Radnitz Cori (1896–1957) who first discovered the process.

When lactic acid is generated in very high levels, not all of it can be moved out into the bloodstream, and some builds up in the muscles. This buildup of lactic acid is generally blamed for muscle soreness and fatigue during exercise. In fact, muscle soreness and fatigue can be caused by many different factors, but lactic acid itself is not the primary culprit. Chapter 3 will go into more detail about muscle soreness and fatigue during exercise. Although the lactic acid itself is not harmful, its buildup can cause the muscles to become acidic.

Muscles usually have a neutral pH, but lactic acid during sprint-type exercise causes pH levels to drop dramatically and thus become more acidic. Luckily, the body has a buffer in the form of **calcium carbonate**, which prevents the muscle from becoming so acidic that the cells die. However, even a moderate drop in pH can prevent some steps in glycolysis from occurring properly. This can slow glycolysis and prevent enough ATP from being made. The acidity is also responsible for the burning sensation during short bouts of strenuous exercise.

The muscle acidity and corresponding drop in how much ATP is available is what most researchers think causes exhaustion during sprinting. The acidity can also prevent calcium from binding properly to tropomysin. Remember that calcium binding to tropomysin is what then allows myosin to bind actin and cause a muscle contraction. When calcium can't bind, less myosin will be able to pull on actin, and the overall contraction will be weaker. Keep in mind that these processes only occur when the muscle is relying exclusively on glycolysis for ATP—during short bouts of intense exercise.

ENERGY USE IN CARDIAC MUSCLE

This chapter deals primarily with how skeletal muscles generate and use energy. However, cardiac and smooth muscles also have characteristic ways of generating ATP. Cardiac muscles resemble slow-twitch skeletal muscles in that they must continuously perform a low-intensity contraction. Each cardiac muscle cell has mitochondria packed between the muscle fibers where they can provide a steady supply of energy. Near the mitochondria, the heart cells also have droplets of fat that the cell uses almost exclusively for energy.

The heart absolutely relies on having a steady blood supply to provide oxygen. During a heart attack, one or more of the blood vessels that bring blood to the heart muscles gets blocked. Even a temporary loss of oxygen can prevent the heart from making enough ATP and contracting normally, cutting off the blood supply to other parts of the body including the brain. The heart is unable to compensate for the lack of oxygen by calling on creatine phosphate stores or by making ATP through glycolysis.

ENERGY USE IN SMOOTH MUSCLE

Smooth muscle has a different way of contracting than either skeletal or heart muscle. In these cells, the myosin heads cycle through ATP about ten times more slowly than the myosin in heart and skeletal muscle. This makes the smooth muscle very slow to contract, but once it is contracted the muscle burns through ATP very slowly and can sustain the contraction for a long time. This quality makes smooth muscle particularly good at squeezing, such as squeezing food through the digestive system or maintaining a blood vessel's diameter.

With its limited ATP needs, smooth muscle has few special adaptations for generating ATP. Like all the cells of the body, smooth muscle cells contain mitochondria that generate ATP from the sugar or fat circulating in the blood. Smooth muscle cells do not store their own energy supplies, nor do they have excess mitochondria. In smooth muscle cells, as in most cells of the body, the mitochondria do tend to congregate near the actin and myosin fibers where the ATP is most needed.

3

Muscular Adaptation to Exercise

It doesn't take an exercise physiologist to tell the difference between a person who works out and a person who doesn't. Those who get regular exercise tend to be leaner and have better defined muscles than those who don't exercise. The lower weight is a function of burning calories while exercising. But other differences, such as stronger, better defined muscles, are a result of molecular changes that take place within the muscle. Some of these changes are visible, but others are invisible adaptations that make the muscles and heart better able to run, bike, row, or lift weights. The muscle fiber type can change, fibers themselves grow larger, muscles contain more ATP-producing enzymes, and the muscles become better able to use oxygen. Remember that there are two types of exercise: anaerobic and aerobic. In aerobic exercise, such as biking, most of the energy comes from oxidative phosphorylation. In anaerobic exercise, such as weightlifting or sprinting, a person cannot deliver enough oxygen to the muscle and so the muscle relies primarily on glycolysis for ATP. Whether a person does aerobic or anaerobic training controls what types of changes take place in the muscle.

MUSCLE ADAPTATIONS TO ANAEROBIC EXERCISE

Gains in Muscle Strength

Regular anaerobic exercise causes muscles to grow larger and more defined. In the past, researchers thought that the increase in muscle size led directly to an increase in strength. Keep in mind that during this time most athletes were men who did indeed develop bulky muscles as they gained strength, and weightlifting champions do have larger muscles than ordinary people. Together, these factors build a compelling case that muscle strength and size are related.

To some extent, these observations do hold true—the men's and women's weightlifting champions do all have extremely large muscles. However, studies in women and children show that more is going on than just an increase in muscle size.

Even in the first eight weeks of training when untrained men and women first begin lifting weights the muscles become much stronger. However, during this time their muscles don't increase noticeably in size. This is especially true of women and children, who can increase their strength the same percentage as men without showing the same increase in muscle girth. From this, it appears that muscle size does contribute to muscle strength but that other factors also play a role in determining a muscle's strength.

Hypertrophy

When a muscle grows large in response to weight training, that gain in size is called **hypertrophy**. Immediately after exercising, a muscle is filled with fluid and feels pumped up. This hypertrophy lasts only a few hours after exercise. True muscle hypertrophy lasts as long as the muscle is in regular use, but decreases if the muscle is not used regularly.

Muscles can grow larger in two ways: The individual muscle fibers themselves can increase in size, or more muscle fibers can be added to the muscle. These two models for hypertrophy are shown in Figure 3.1. Both seem to play a role in how muscles increase in size. Whether there are more fibers or simply larger fibers, there are more individual actin and myosin filaments and more cross-bridges pulling those filaments past each other, which means a stronger contraction.

INCREASES IN MUSCLE FIBER SIZE

For a long time scientists thought that people were born with a certain number of muscle fibers in each muscle and that this number could not change. If that were the case, then all muscle hypertrophy would be due to increases in fiber size because no new fibers could be formed. One compelling reason to believe this is that scientists can look under a microscope and see larger individual muscle fibers in a person who has trained on weights for many weeks compared to that same person before they began training. The individual fibers grew larger, leading to an overall larger muscle.

The addition of new actin and myosin filaments within a muscle fiber happens because the muscles produce new protein. At any given time, a muscle is both building new protein and breaking down old protein. The balance between these processes keeps a muscle at a certain size. Regular

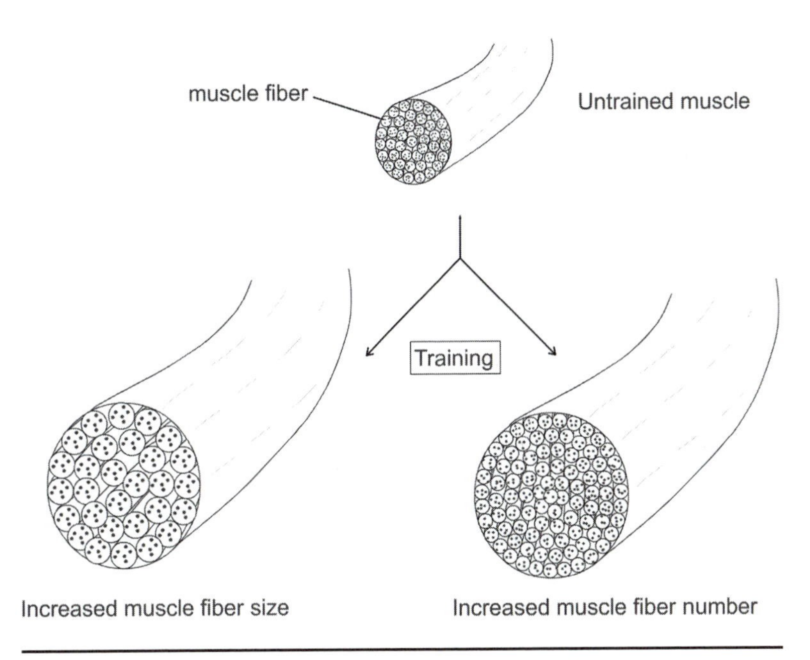

Figure 3.1. Two models for muscle hypertrophy.

strength training increases the amount of new protein that muscles make and decreases the amount of protein that is broken down, leading to a net gain in muscle mass.

This balance between protein being made and being broken down is also altered by the hormone **testosterone**, which men have much more of than women. Testosterone increases the amount of protein that muscles make. This accounts, in part, for why men develop larger muscles then women do. (See "Anabolic Steroids" for a discussion of unnatural muscle growth.)

Some experiments make scientists think that endurance athletes rely exclusively on increases in muscle fiber size to gain their increases in strength, while sprint athletes rely very little on increases in fiber size. When researchers train animals to press a lever several times to receive food, those animals have the same number of muscle fibers before and after training but those fibers grew larger. However, they mainly see this effect in animals who are trained to press light weights and must repeat the movement many times throughout the day—much like endurance training.

INCREASES IN MUSCLE FIBER NUMBER

Increases in muscle fiber number seem to happen mainly in athletes who train on heavy weights. When cats were trained to press a very heavy lever to get their food, researchers saw muscle fibers in the process of splitting in

Anabolic Steroids

Any sport that requires strength, such as weightlifting, discus throwing, sprinting, or baseball, has athletes that are looking for ways to increase their muscle mass. One common—though in many sports illegal—way to quickly increase strength is by taking a group of drugs called **anabolic steroids**. Testosterone, which increases muscle mass naturally, is one type of anabolic steroid, though the most commonly taken steroids have slight chemical modifications to make them different from testosterone. Some of these modifications make the drugs harder to detect in blood tests.

Both male and female athletes who take steroids develop much larger muscles and also have considerable increases in strength. The athletes also appear bulkier and have a lower percentage of fat. These sought-after effects increase with increasing steroid dose, which means that those athletes who take steroids tend to be on a high dose. Using steroids also speeds recovery, allowing athletes to do rigorous workouts on successive days, and some athletes think steroids increase their endurance as well. Because of the unfair advantage that steroids bring to those athletes, officials in many events require that athletes be tested for steroid use.

In addition to getting a person banned from competition, steroids pose serious health risks. In women, steroids can induce masculine traits such as a deeper voice or facial hair and can interfere with the woman's menstrual cycle. In men, the drugs can cause the testes to atrophy and lower sperm count. The body also compensates for increased steroids by decreasing the production of testosterone. This can interfere with the man's ability to have an erection and can cause the man to develop breasts. In both men and women, chronic steroid use can lead to acne, increased risk of heart disease, liver damage, and a more aggressive personality (commonly called "'roid rage").

Steroid use was once considered a problem only for adult professional athletes, but with increased pressure on high school and even junior high school athletes, steroids are making their way into school locker rooms. Natural steroids orchestrate many of the changes that take place through puberty. By adding anabolic steroids, the young athletes run the risk of altering their normal hormonal changes and the development of the reproductive organs. The athletes may also stunt their growth. This happens because the growing tip of the bone continues producing new bone throughout puberty, then closes off when a person is done growing. Anabolic steroids can cause the bones to close off early and permanently stop bone growth. With the serious long-term effects steroids can have on young people, more high school coaches are educating athletes to avoid the drugs.

two, generating new fibers. They also counted the muscle fibers and found more individual fibers after training than before, probably as a result of fibers splitting to create additional fibers.

Although it is hard to count the exact number of fibers in a person's muscle, researchers can look at the overall size of muscle fibers. When they compare muscle fibers in bodybuilders and people who are fit but haven't done weight training, they find that the muscle fibers are about the same size. Because the bodybuilder's overall muscles are much larger, the researchers conclude that the bodybuilders must have more fibers in total.

One reason that strength training increases the number of muscle fibers may have to do with how the muscles respond to lifting heavy weights. After a heavy weightlifting session, muscle fibers have some damage due to the stress of lifting heavy weights. In the process of repairing those damaged cells, new fibers can be formed.

Strength Increases Due to Exercise

Scientists aren't precisely sure how muscles become stronger without becoming bigger, such as in the case of women or children or during the early weeks of training in men. Several studies have led exercise physiologists to believe that changes in the nervous system may explain these increases in strength. In one study, participants did strength training exercises with one arm for eight weeks. At the end of the study, the participants had grown 25 percent stronger in the trained arm, as expected. But these people were also about 15 percent stronger in the untrained arm. This result tells researchers that factors in addition to muscle strength must have changed in response to exercise.

One explanation is that when the muscle of a trained person contracts, more motor units contribute to the contraction. Because a person is using the entire muscle, he or she can lift a heavier weight. This could also explain how people pull off remarkable feats of strength when they are under pressure, such as lifting people out from under cars or freeing themselves when pinned under heavy objects. The muscle may contract all motor units at one time, making the muscle significantly stronger than when only a few motor units contribute to a contraction.

Another explanation could be that when a trained muscle flexes, the opposite muscle doesn't resist as much. For example, when a trained person does an arm curl using the biceps muscle, the opposing triceps muscle relaxes and allows the biceps to flex (see Chapter 1 for a discussion of how muscles work in antagonistic pairs). According to this theory, the triceps muscle in an untrained person does not relax as much, so it takes more strength on the part of the biceps to overcome the resistance. If this explanation were true, then training with one arm causes the brain to alter how it instructs the triceps of both arms.

Scientists have also seen changes in the junction between the nerves and the muscle in people who are highly trained. Although they still don't know how those changes relate to muscle strength, it is possible that these changes allow the muscle to contract more strongly in response to a given nerve signal.

At this time, scientists don't have enough evidence to figure out which explanation is right. It could be that all of these changes take place to some extent, each contributing to the overall gain in strength.

Changes in Fiber Type

As discussed, muscles are made up of two general types of muscle fibers: fast-twitch and slow-twitch. The slow-twitch fibers can be further broken down into types IIa and IIb. The type IIa fibers fall somewhere between slow-twitch fibers and type IIb fast-twitch fibers in terms of their ability to use oxygen and carry out glycolysis (see Chapter 2 for a review of how muscles generate energy anaerobically). People who have predominantly fast-twitch fibers tend to be good sprinters, while those with predominantly slow-twitch fibers tend to be good endurance athletes.

The question for trainers is whether the right type of exercise can build up a particular fiber type or if athletes must learn to make the best of their inborn fiber composition.

It turns out that to a large extent, an athlete will always have a genetic tendency toward one type of muscle fiber and therefore toward one type of exercise. A person whose parents were both sprinters is unlikely to develop into a marathon runner and vice versa. However, with exercise some fast-twitch fibers can become more able to use oxygen and slow-twitch fibers can become more able to make energy in the absence of oxygen using anaerobic respiration. Of all the muscle fiber types, fast-twitch type IIa fibers seem most able to become either type IIb or slow-twitch fibers.

In one experiment, researchers had participants engage in a twenty-week weightlifting regimen. At the end of this time, researchers noticed more type IIb fibers and fewer type IIa fibers in the athletes' muscles. These converted fibers contribute to the anaerobic capacity of that muscle and help the athlete at short, strenuous, sprint-type exercise. At the same time those fibers can no longer assist in endurance exercise, reducing the athlete's endurance in aerobic exercise.

Changes in Muscle Enzymes

Regular training increases the amount of muscle enzymes that are available to produce ATP. The type of exercise a person does determines which types of enzymes increase. People who do weight repetitions or short sprint intervals build up enzymes that are involved in the glycolysis pathway, that convert pyruvate into lactic acid, and that recharge ATP though creatine phosphate. These changes allow the muscle to use anaerobic respiration at a faster rate than in untrained people. With enough ATP being made, sprint-trained athletes can lift a weight for more repetitions or can maintain a faster pace during a sprint race. However, sprint training does not increase enzymes involved in the citric acid cycle. A person who lifts weights exclusively will not be able to improve his or her 6-mile (10-kilometer) running time, nor will a runner increase his or her strength without weight training.

Changes in Muscle Buffering

Remember that during anaerobic activity, muscles generate ATP though glycolosis and convert the end product of glycolosis into lactate. This lactate is acidic and can lower the pH of the muscle to a point where the muscle can no longer contract. The acidity is part of what makes a muscle feel fatigued immediately after sprint-type activity.

All muscles contain some chemicals (called buffers) that help the muscle maintain a normal pH. However, muscles that have had at least eight weeks of anaerobic training can buffer 12 to 50 percent more acidity than untrained muscles. Athletes with anaerobically trained muscles can sprint for a longer time without giving in to fatigue due to acidity than untrained athletes.

Given the advantages of buffering the acidity in the muscles, some athletes have begun taking buffering supplements. Although the idea of buffering the muscles with supplements seems sound, exercise physiologists have found no proof that the supplements enter into the muscles where they can help buffer acidity from lactic acid. In trials, trainers have not detected any difference between athletes who have or have not taken buffering supplements.

MUSCLE ADAPTATIONS TO AEROBIC EXERCISE

Increased Ability to Use Oxygen

Whereas anaerobic exercise such as weightlifting or sprinting causes the muscles to grow larger, aerobic exercise causes changes in the way the muscle uses energy. Keep in mind that aerobic exercise such as swimming or rowing doesn't use a muscle's full strength, but that muscle must be able to contract repeatedly over a long time. The muscles will become stronger and larger than in a person who does not work out at all, but they never achieve the size or strength of a weightlifter's muscles. The hurdle for endurance athletes is delivering enough oxygen to the muscles so the muscles can make ATP for contraction.

BLOOD SUPPLY

Capillaries run throughout the muscle fibers, delivering oxygen and nutrients to the muscles. As people train for endurance events, their muscles accumulate more capillaries to meet the energy needs. Over time, trained people can have as many as 15 percent more capillaries in their muscles than untrained people.

MYOGLOBIN

Remember that myoglobin in the muscle takes oxygen from the blood and delivers it to the mitochondria. An increase in capillaries won't do the muscle any good unless that muscle contains enough myoglobin to carry the increase in oxygen. The slow-twitch fibers that are most commonly used for endurance activities already contain more myoglobin than fast-twitch muscles. With training, slow-twitch fibers will contain as much as 75 percent more myoglobin than untrained fibers.

MITOCHONDRIA

The mitochondria produce ATP for the cell, using oxygen to manufacture ATP and water. The amount of ATP that can be made in a muscle cell depends, in part, on how many mitochondria that cell contains. As people do aerobic training, both the number of mitochondria within their slow-twitch muscle fibers and the size of those mitochondria increase.

Fast-twitch type IIa fibers normally contain some mitochondria, though not as many as in slow-twitch fibers. With aerobic exercise these type IIa muscle fibers also accumulate more, larger mitochondria. This increase helps those fast-twitch type IIa fibers contribute to the aerobic exercise.

ENZYMES

Within the muscle cells, sugars go through glycolysis, which produces pyruvate. This pyruvate is converted into acetyl CoA, which then enters the mitochondria where the citric acid cycle uses that acetyl CoA to generate NADH and $FADH_2$, which eventually leads to ATP production. The end result—producing ATP—relies on the citric acid cycle working at optimal speed. To make sure the cycle can produce enough NADH and $FADH_2$ to meet a muscle fiber's energy needs, muscles in trained athletes contain more of the enzymes that are used during the citric acid cycle.

The activity of these enzymes increases throughout aerobic training. People who get even a small amount of aerobic exercise have much more enzyme activity in their muscles than in people who do not train. Highly trained people can have as much as double the citric acid cycle enzyme activity of untrained people.

Measuring Oxygen Use

The increase in capillaries, myoglobin, mitochondria, and enzymes all adds up to muscles that are better able to use oxygen and can therefore use more oxygen than muscles in untrained people. An athlete's increased ability to use oxygen is one key factor that exercise physiologists and trainers use as a measure of physical fitness. The maximum oxygen use is often called a person's **VO_2max** (measured in milliliters of oxygen used per kilogram of body weight per minute; pronounced Vee Oh Two Max).

When a person does aerobic activity, his or her breathing increases to bring more oxygen to the muscles. To determine how much of that oxygen the muscles actually use, researchers have an athlete exercise on a stationary bike or a treadmill. They then put a device over the athlete's nose and mouth to measure the amount of oxygen in the air that a person breathes in and compare it to the amount of oxygen that a person breathes out. This information is fed into a computer, which keeps track of the oxygen use as the athlete continues to exercise harder.

From this test, researchers find that as people increase their work load, they also increase the amount of oxygen they use, but only up until a certain point—the VO_2max. After the athlete reaches the VO_2max, his or her muscles are not capable of taking up any more oxygen. Even though the athlete is working harder, the amount of oxygen they use remains the same. If the athlete needs to produce additional ATP, their muscles must rely on anaerobic respiration for that increase. Because this process is less efficient at producing ATP than aerobic respiration and results in a buildup of lactic acid, an athlete tires quickly after exceeding the VO_2max.

People who are untrained have a very low VO_2max. Their muscles are not efficient at using oxygen to make energy, so they rely on aerobic respiration for quite a bit of their ATP and tire quickly. As people train and their muscles become better adapted to use oxygen, their VO_2max can as much as double. This increase is a direct result of the increased blood flow through trained muscles, increased myoglobin to take up oxygen, and increased mitochondria to oxygen for making ATP. These changes occur through training, but at some point a person has all the blood vessels, myoglobin, and mitochondria that their muscles can support. At this point, VO_2max no longer increases even with increased training. Any increases in athletic performance come from other physiological changes.

Sedentary people between the ages of 20 to 29 usually have a VO_2max of about 35 for women and 45 for men. That value goes down about three to four points per year as a person ages. Male runners have a VO_2max of about 60 to 85, while women tend to be about ten points lower. Cross-country skiers top the VO_2max charts, with values as high as 94 in men and 75 in women. These extremely high values are in part a testimony to the value of training but also represent a person's genetic abilities to use oxygen. (See "Blood Doping to Improve Performance" for a discussion of unhealthy ways to increase VO_2max.)

Although VO_2max is a good indicator of a person's athletic capabilities, it does not always predict performance. A person with a lower VO_2max can

Blood Doping to Improve Performance

As with steroids in anaerobic events, aerobic sports are riddled with their own set of illegal performance-enhancing drugs. Most of these improve the athlete's ability to use oxygen, usually by increasing the number of oxygen-carrying red blood cells in the athlete's blood (also called **blood doping**). So far, this chapter has discussed muscular changes that increase oxygen use. But the muscles can only use as much oxygen as the blood brings. By increasing the amount of red blood cells, the athletes can deliver more oxygen to their muscles and therefore make more ATP.

The oldest form of blood doping is to simply take blood from an athlete, then reinject that blood several weeks later. In the 1980s, a study in long-distance runners found that this approach could increase the runner's VO_2max by about 5 percent and increase the runner's time to exhaustion on a treadmill by about 40 percent. Other studies have also found that distance runners and cross-country skiers both improve their times after blood doping.

A newer form of blood doping is the drug erythropoietin, better known by the name EPO. Kidneys normally produce erythropoietin to signal red blood cell production. By injecting EPO, athletes encourage additional red blood cells to form, giving the same effect as if they had simply injected more of their own blood. EPO is widely used by doctors to help anemic people who need additional blood cells. In athletes, the drug increases VO_2max and time to exhaustion on a treadmill by almost exactly the same amount as injecting the athlete's own blood. Because of how easy it is to inject EPO, it is now widely used in endurance events, particularly in biking events such as the grueling Tour de France.

With both forms of blood doping, athletes risk having their blood become so thick with red blood cells that it cannot pass through the capillaries. This viscous blood can also cause blood clots or heart disease. In addition to being dangerous, drugs that contribute to blood doping are illegal in most sports, and athletes are widely screened for their use.

still outcompete someone with a very high VO_2max. Part of this difference has to do with how long a person can exercise at near their VO_2max.

No athlete can exercise at his or her VO_2max for long. Instead, a person running a marathon or biking a 100-mile race competes at as fast a pace as he or she can maintain for the duration of the event. A person with a low VO_2max who can maintain a steady pace while at 80 percent of his or her VO_2max can outcompete an athlete with a high VO_2max who can only maintain a pace at 70 percent of their VO_2max. (See "Training at Altitude Improves VO_2max" for ways to manipulate oxygen.)

Measuring Lactate Threshold

The **lactate threshold** is another measurement of a person's endurance abilities. When a person is exercising at a low level, such as walking, he or

Training at Altitude Improves VO_2max

At high altitude the air carries much less oxygen than at low altitude, so with each breath a person brings in less oxygen and delivers less oxygen to the muscles. It is this change in oxygen that makes walking up stairs seem like a tough workout when at high altitudes. After a few days a person's body will produce extra red blood cells to take advantage of what little oxygen the air contains, and soon an athlete can work out normally without feeling the effects of altitude.

Many athletes have realized that training at a high altitude offers all the benefits of blood doping—more red blood cells delivering oxygen to the muscles—as illegal methods. It is because of this thinking that a major Olympic training center resides in Colorado Springs, Colorado, at about 6,000 feet above sea level. Athletes who train at this facility will develop extra red blood cells that will help them use oxygen more efficiently in competition at sea level.

In recent years, however, some coaches have noticed that the athletes' muscles don't develop as well during workouts at high elevation. The problem is that even with more red blood cells, the muscles are still not getting enough oxygen for a maximal workout. The athletes are essentially not training their muscles as hard as an athlete working out at sea level. Their muscles don't become as adapted for endurance use—don't develop the added enzymes and capillaries, don't store as much glycogen and fat, and don't recruit as many type IIa fibers to help out the slow-twitch fibers—so even with extra red blood cells, the athletes may not compete as well at sea level as their coaches had hoped.

Some coaches have now started a "live high, train low" training regime. The athletes live at high altitudes where they develop excess red blood cells, but they go to lower altitudes to train. In this approach, the athletes get both excellent blood-carrying capacity and all the muscular changes associated with hard training. In studies in the 1990s, physiologists saw a 3–5 percent improvement in athletes' speed and endurance after the athletes lived at high altitudes but trained at low altitudes.

While driving up and down a mountain every day doesn't appeal to all athletes, the effects of this type of training do. With this in mind, companies have started producing tents, sealed rooms, and even sealed houses in which athletes can simulate a high altitude. These are now being used by cross-country skiers in places such as Finland where high-elevation mountains are hard to come by. They are also being used increasingly by runners, bikers, and other athletes in the United States.

For now, simulating high altitude is legal because the body produces natural compounds (including EPO) to raise red blood cell numbers. The body is better able to regulate how many red blood cells get made and not produce so many that the blood becomes viscous.

she has very little lactic acid in the blood. Very little lactic acid is produced in the person's muscles, and what lactic acid is produced the body can quickly clear. As a person starts exercising faster, the muscles will slowly start to rely more and more on anaerobic respiration to produce enough ATP. The resulting lactic acid will build up in the muscles and eventually also build up in the blood.

This onset of lactic acid buildup in the blood occurs quite suddenly—it does not slowly increase with exercise. The point where lactic acid begins to accumulate is called the lactate threshold. Scientists aren't sure what triggers lactic acid to begin accumulating, but when athletes cross that line they become tired and less able to maintain their exercise pace. For long-distance exercise, athletes must maintain a pace that is below their lactate threshold to avoid tiring early in the race.

Researchers usually measure the lactate threshold in terms of the percentage of that person's VO_2max. Untrained people reach their lactate threshold at about 50 percent of their VO_2max, while trained athletes can reach as much as 80 percent of their VO_2max without crossing their lactate threshold. Athletes with a higher lactate threshold can maintain a faster race pace than an athlete with a lower lactate threshold, even if the two athletes have the same VO_2max.

Coaches and exercise physiologists still debate about whether VO_2max or lactate threshold is a better indicator of an athlete's physical condition. However, because measuring VO_2max requires time and expensive equipment, many coaches of serious athletes will measure the athlete's lactate accumulation as a way of measuring overall fitness. This measurement is an approximation of the lactate threshold but requires less time and equipment.

The coach will usually have the athlete do a short bout of exercise—such as a 200 meter swim—at a prescribed pace, then measure the athlete's lactic acid accumulation at the end of that event. Throughout the season, the coach can repeat that test. In such trials, the athletes accumulate less lactic acid at the same exercise pace and distance as the season progresses, and they accumulate the least lactic acid around the time that they produce their best performance.

Scientists still don't know which changes in the muscle increase the lactate threshold. A combination of having more slow-twitch muscle fibers, more mitochondria, and more ATP-producing enzymes seems to reduce the muscle's reliance on anaerobic respiration. This in turn reduces the amount of lactic acid that accumulates. Although coaches don't know what exactly causes a higher lactate threshold, they do use training strategies that have been shown in the past to increase the lactate threshold. These include interval training with short rest, or longer workouts at or above the lactate threshold followed by a long recovery period.

Changes in Fiber Type

As with anaerobic activity, even intense aerobic training won't make significant changes to a person's muscle fiber composition. However, endurance training can cause some increases in the amount of slow-twitch fibers—mainly due to type IIa fibers converting to slow-twitch fibers—and also seems to cause some fast-twitch type IIb fibers to take on characteristics of type IIa fibers. Type IIa fibers in endurance-trained people also tend to have a greater ability to use oxygen than in untrained people.

MUSCLE SORENESS AFTER EXERCISE

Any person who has tried a new exercise for the first time is familiar with the feelings of muscle stiffness and pain that can last for several days. This pain usually goes along with a feeling of weakness. Over time, the muscles adapt to the new exercise and stop hurting. There are two phases to this muscle pain: **Acute muscle soreness** is the burning sensation right after exercise, whereas **delayed onset muscle soreness** is the pain that can last several days.

Acute Muscle Soreness

Acute muscle soreness is caused primarily by the acidity from lactic acid in the muscles. Recall that lactic acid builds up in muscles after sprint-type exercise that relies solely on anaerobic respiration for ATP. Although the lactic acid does not damage the muscles, the acidity causes a burning sensation. This goes away almost immediately after a person stops exercising.

Muscles also swell slightly during all forms of exercise. This swelling, also called edema, is due to fluid in the blood plasma seeping into the muscle. The swelling can be uncomfortable but goes away within a few hours of exercise.

Delayed Onset Muscle Soreness

Delayed onset muscle soreness, as its name implies, begins long after the acute soreness ends. It is commonly attributed to lactic acid buildup, though it is usually a result of damage to the muscle and other factors rather than lactic acid.

MUSCLE DAMAGE

After strenuous exercise—either the first time exercising a muscle group or significantly increasing the amount of use a muscle gets—the membranes surrounding muscle fibers can rupture and allow fluids from the fiber to leak out. Another form of damage can occur when the actin filaments, which are normally anchored in proteins at the Z disk, pull free from their anchoring and rupture the Z disk. Again, the pattern of actin and myosin is disrupted.

Both of these forms of damage are more likely to occur during eccentric action—the type of muscle action that occurs when the muscle contracts as the two attachment sites move farther apart. The easiest way to think about eccentric action is holding a weight with a bent arm. When letting the arm straighten the biceps muscle is lengthening, but at the same time the biceps is contracted to control how quickly the arm straightens out. This same type of action occurs in runners running downhill. In many experiments, researchers have found that runners on a downhill incline on a treadmill develop more muscle pain and damage than those on a flat or uphill treadmill.

INFLAMMATION

After the muscle becomes damaged, the muscle cells release chemicals into the bloodstream that attract immune cells. These immune cells break down dying muscle fibers and clean up the debris. Although this process is necessary in order for the muscle to recover, it does also cause the muscle to become inflamed and seems to add to the soreness.

WEAKNESS

People who experience muscle soreness after rigorous exercise will notice that the pain goes away after a few days, but they still feel weak and tired for up to a week after the initial exercise. Part of this weakness is due to the muscle damage itself—with many of the muscle fibers damaged, there are fewer fibers to generate a contraction. At the same time, muscles store less glycogen while the fibers are being repaired. Without glycogen, the muscle has less fuel to produce ATP, and the muscle has less ability to contract.

TRAINING TO PREVENT PAIN

When a person begins doing regular exercise, the muscles are untrained and will invariably hurt after the initial training sessions. Trainers have taken two different approaches to minimize this pain. One approach is to build up exercise slowly over a long time. This will keep muscle damage—and therefore pain—to a minimum. However, with this training approach a person may feel some pain for several weeks as the muscles adjust to training.

Another approach is to begin training with one large burst of exercise. This causes considerable pain after the initial training session, but the muscles will recover to be much stronger and won't hurt after subsequent exercise. The merits of these two approaches depend on the athlete as well as on the training goals.

Athletes training for a running race who want to avoid pain and muscle damage after the race can follow some precautions. Because most damage comes from eccentric motions, runners can train with extreme eccentric motion by running down hills. The muscles will adjust to this eccentric training so they will become less damaged over the course of a normal race.

Chapter 7 has more information about strategies to prevent muscle soreness and injury.

Repairing Muscle Damage

Throughout each muscle are undeveloped cells called **satellite cells**. These are a form of stem cell that are specially adapted to replace damaged muscle fibers. Unlike embryonic stem cells, these cells have matured enough to function only as muscle cells—they cannot become skin cells, nerve cells, or any other type of cell. However, they are still able to develop into either slow- or fast-twitch muscles, depending on what cell type is needed.

When a muscle sustains damage through overuse, it releases chemicals that signal these satellite cells to divide. Of the two new cells, one remains a satellite cell while the other fuses with the damaged muscle cell and contributes to its repair. If enough fibers are damaged and need repair, new fibers formed from satellite cells could end up contributing to the muscle size. Some researchers think that damage repair by satellite cells is required in order for weightlifters to bulk up.

The entire process of the satellite cell receiving a signal and forming a new muscle fiber can take several days, which is one reason why it takes a while to recover muscle strength after damage-inducing exercise such as running a marathon or a heavy weightlifting session.

CHANGES TO HEART MUSCLE THROUGH EXERCISE

With all the changes that take place to skeletal muscle as a result of training, it is easy to overlook the heart muscle. However, the heart plays a key role in a person's exercise performance and also undergoes some changes through exercise. These changes help the heart pump more blood to the muscles. The regular blood supply is critical—the blood brings oxygen and fuel to the muscle and clears waste and lactic acid. Without sufficient blood flow, the muscles would not be able to maintain their peak contraction.

Heart Size

Through regular training the heart muscle becomes larger, as do skeletal muscles. The heart is divided into four chambers, of which one (called the left ventricle) is primarily responsible for pumping blood containing oxygen out to the body. The muscles surrounding the left ventricle grow larger in endurance-trained athletes and the chamber cavity grows larger to accommodate more blood. Resistance-trained athletes also have some increase in left ventricle muscle thickness, though the chamber size remains about the same as in untrained people. The thickness of a person's heart muscle directly correlates to their VO_2max—endurance-trained athletes with a high

VO_2max also have thicker heart muscle walls, whereas sedentary people with a low VO_2max have thinner heart muscles.

A thicker heart muscle means a stronger contraction, pushing blood out to the muscles where it is needed. Many athletes notice this change in heart muscle strength by a lower resting heart rate. Because the heart muscle is stronger, it pumps more blood with each contraction. At rest, the heart can beat fewer times per minute and still distribute the same amount of blood as a weaker heart pumping at a fast rate. Untrained people often have a resting heart rate of between 65 to 80 beats per minute, though extremely sedentary people can have heart rates as high as 100 beats per minute. Trained athletes have been known to have resting heart rates as low as 28 to 40 beats per minute.

The larger chamber size also contributes to an athlete's lower heart rate. Trained athletes have more blood volume than sedentary people, so the chamber fills fuller per beat than in untrained people. The chamber also has more time to fill between the slower beats, further increasing the blood volume in the heart chamber. With these factors combined, athletes force much more blood out to the body with each beat than untrained people. In one study, the total volume of blood forced out by each heartbeat went up by almost 50 percent after a six-month training regime.

Heart Rate during Exercise

During exercise, the heart rate increases to send more blood to the muscles. When an athlete reaches a steady pace, such as the pace for a two-hour bike ride, the heart rate levels off and will stay about the same for the given amount of exercise. This heart rate is called the steady-state heart rate. If the athlete increases or decreases the pace, the steady-state heart rate will also increase or decrease.

People who are in good physical condition will generally have a lower steady-state heart rate for the same amount of exercise as a person who is untrained. This lower heart rate is a result of the heart squeezing out more blood volume per heartbeat in trained people.

For all-out exertion, that steady-state heart rate will reach a maximum that the heart cannot exceed. A person can reach the highest maximum heart rate at about age 10 to 15, and that maximum decreases by about one beat per minute per year. A person's maximum heart rate can be approximated with the formula 220 – Age (in years). So, a 40-year-old person would have a maximum heart rate of $220 - 40 = 180$, or 180 beats per minute. This formula is simply an estimate—many people have heart rates that fall outside the calculated maximum. Figure 3.2 shows how the maximum heart rate decreases with age.

The decrease in maximum heart rate seems to happen as a result of how the signal to contract spreads across the heart. Even with regular exercise, which causes the muscle to contract with more force, a person's maximum

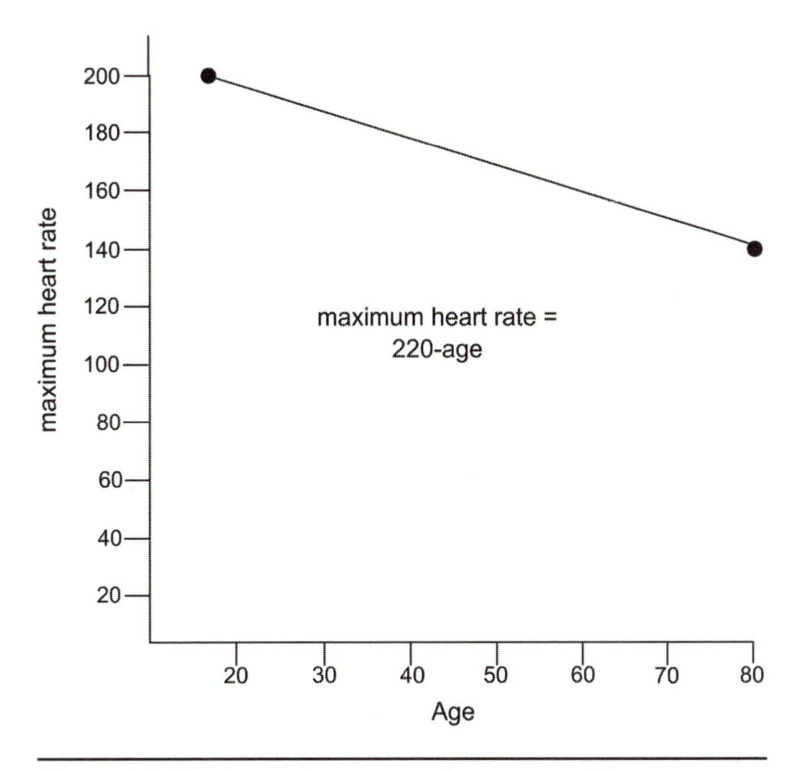

Figure 3.2. Maximum heart rate decreases with age.

heart rate will decrease with age. This doesn't generally affect a person's athletic performance for endurance-type events, because in these events the heart rate rarely reaches its maximum. However, as people age they may notice their sprint performance declining. People reach their maximum heart rate when sprinting—if the maximum heart rate is lower, a person will tire faster when sprinting.

CHANGES TO SMOOTH MUSCLE THROUGH EXERCISE

When a person is resting in a chair, the muscles receive about 20 percent of the total blood flow, with the liver, kidney, and brain making up about 60 percent of the blood flow with the remaining blood distributed through the heart muscle, skin, and other tissues. During heavy exercise, the heart muscle continues to receive about 5 percent of the blood flow while the skeletal muscle receives 80 to 85 percent of the total blood flow and the other organs receive severely restricted blood flow. This massive redistribution of blood arises due to changes in the smooth muscle lining that surrounds blood vessels.

Blood leaves the heart through large vessels called arteries. These arter-

Heat Fatigue

Anyone who has tried to run on a hot sunny day has learned that they fatigue much easier in the heat than on a cool day. The reason for this has to do with how blood is distributed under hot conditions. When the body overheats, smooth muscle around capillaries in the skin expands so that blood can bring heat from inside the body to the surface where it cools off. This cooler blood then continues circulating, helping to cool the body. By redirecting blood to the skin, less blood is available for the muscles. The muscles then get less oxygen, can't produce as much ATP, and get fatigued. This effect is one reason why running races often start early in the morning, particularly in the summer when midday heat could cause runners to overheat and develop heat stroke or heat exhaustion.

ies branch out, sending offshoots to all regions of the body. When the blood reaches a tissue, such as a muscle or the kidneys, the arteries branch into a network of tiny vessels called capillaries. These capillaries then feed into veins that return blood from the organs back to the heart. All of these vessels are surrounded by sheets of smooth muscle that regulate the size of the vessel. When the smooth muscles contract, the vessels become narrow and carry less blood.

Several signals direct when muscles surrounding blood vessels should contract. One of these is the level of oxygen in the surrounding tissue. Remember that muscles use up more oxygen as they increase their demand for ATP. Where oxygen levels are low, blood vessels open up to allow more blood into the region to supply fresh oxygen and fuel.

At the beginning of exercise, signals throughout the body cause smooth muscle surrounding capillaries in the digestive system and other organs to constrict, reducing blood flow. At the same time, signals such as lower oxygen levels instruct capillaries in the muscle to expand, allowing more blood to the region. As exercise continues, more and more blood is redirected to the muscles. (See "Heat Fatigue" about blood distribution and hot weather.)

The redistribution of blood to the muscles is dependent, in part, on the blood supply needs of other tissues. For example, exercise too soon after a meal can result in not enough blood being directed to the muscles. In studies in both humans and other animals, eating a meal directly before exercise caused a 15 to 20 percent drop in blood flow to the muscles. This blood was redirected to the intestinal tract where it was needed to help digest food. Many athletes avoid eating soon before exercise in order to have as much blood flow as possible available to the muscles.

4

Development of the Muscular System

Muscles are one of the earliest organ systems to begin forming in a developing embryo, with the heart muscle leading the way and beating within a few short weeks after fertilization. From these earliest stages, the smooth, heart, and skeletal muscles undergo an amazing transformation from a group of disorganized cells in the embryo to eventually become coordinated muscles with intricate structures and interconnections with growing bones and nerves. Throughout life the muscles continue to change, becoming increasingly coordinated as a child matures, responding to a person's exercise levels and repairing damage in adulthood, and slowly degenerating with age. These changes are all controlled by chemical signals in the body and physical signals from surrounding muscle cells and from the nerves.

Using what they have learned about how muscles grow and develop, some researchers are now trying to find ways to maintain the muscle in older people whose muscles would ordinarily get smaller and weaker. This kind of work could also lead to ways for astronauts keep their muscles strong in space, where low gravity normally causes muscles to get weaker.

EMBRYONIC SKELETAL MUSCLE DEVELOPMENT

Creating the Embryo

During the first several weeks after fertilization, the developing embryo goes through rapid and dramatic changes. Soon after the sperm and egg fuse,

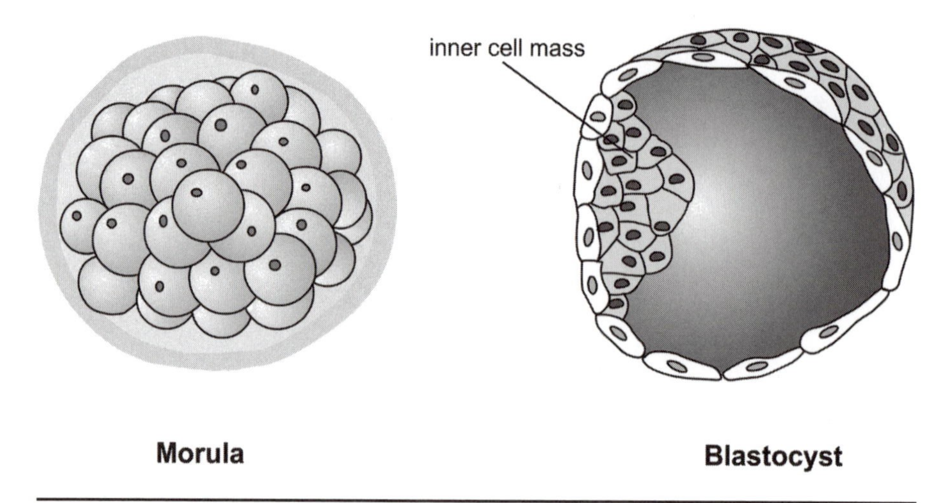

Figure 4.1. Morula and blastocyst showing the inner cell mass.

the fertilized egg divides rapidly to form a solid ball of cells called the morula. At about day five in humans, this ball undergoes a change, and a hollow cavity begins to form inside the mass of cells. Eventually, the cells form a balloon-like hollow structure called a blastocyst, with a single layer of cells lining the inner cavity. Figure 4.1 shows the morula and the hollow blastocyst.

Humans and other animals are clearly more complex than a single layer of cells surrounding a hollow center—our many cell layers, including muscles, are formed when the blastocyst undergoes a process called gastrulation. In gastrulation, which takes place at about the ninth day after fertilization in humans, cells at one side of the blastocyst form an indent, much like poking a finger into a balloon. The cells at the tip of this indent crawl inward, pulling the rest of the cell layer along with it. While this indent is forming, some cells near the tip break free from the single-cell layer and move on their own to fill the hollow space between the outer cell layer and the growing indent.

Eventually, gastrulation results in an outer layer of cells from the original blastocyst, an inner layer of cells formed from the indent, and a middle group of cells made up of those that broke free during gastrulation and filled the space between the two layers. The outer cell layer is called the **ectoderm** and goes on to form the skin, nervous system, and eyes. The inner cell layer surrounding the hollow core is called the **endoderm**. These cells go on to form the entire digestive system including the many glands that release digestive fluids into the stomach and intestines. The hollow tube that is formed by the indenting endoderm cells eventually becomes the hollow core of the digestive system. All the other organs including the heart, lungs,

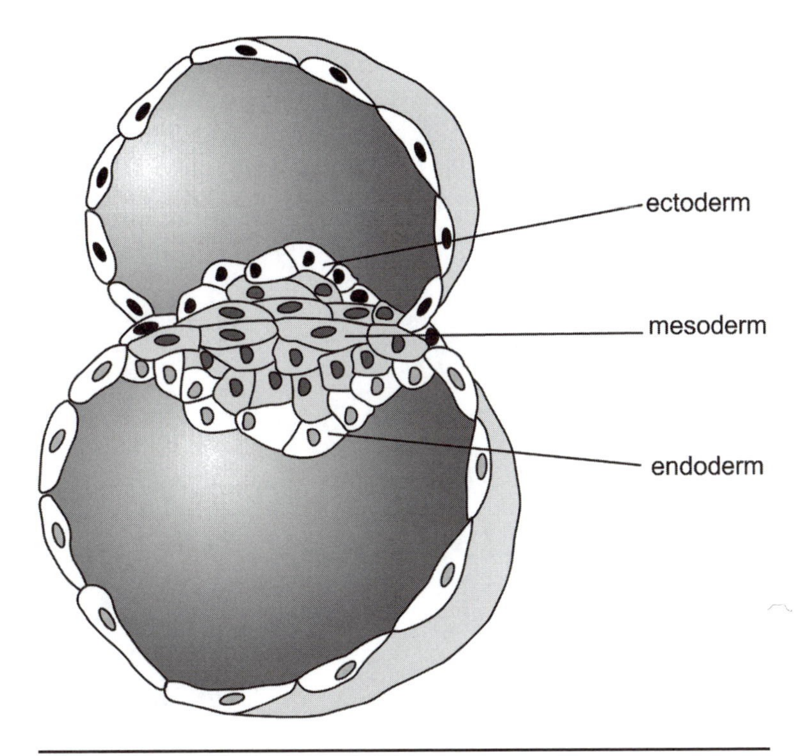

Figure 4.2. Location of the three cell layers in the blastocyst.

liver, kidneys, skeletal muscles, and smooth muscles are formed from the cells that broke free from the indenting endoderm to fill the middle space. These are called the **mesoderm**. Figure 4.2 shows the three cell layers in a blastocyst.

Forming Muscle Precursors

The cells that make up the mesoderm start out as disorganized tissue, but some of those cells quickly begin taking on the form of recognizable human tissues. One of the earliest structures is an early spine-like tissue, called the **notochord**, running down the back of the embryo. Beneath the notochord is the developing spinal column, called the **neural tube**. On either side of the notochord some mesoderm cells break into paired structures called **somites**, with one somite on each side of where the spinal column will eventually form. All skeletal muscles of the body and some from the head come from these somites. Most skeletal muscles in the head come from mesoderm cells that did not form somites. At the same time as when the somites are forming, the embryo develops a visible head region and begins to form small buds that will eventually form the arms and legs.

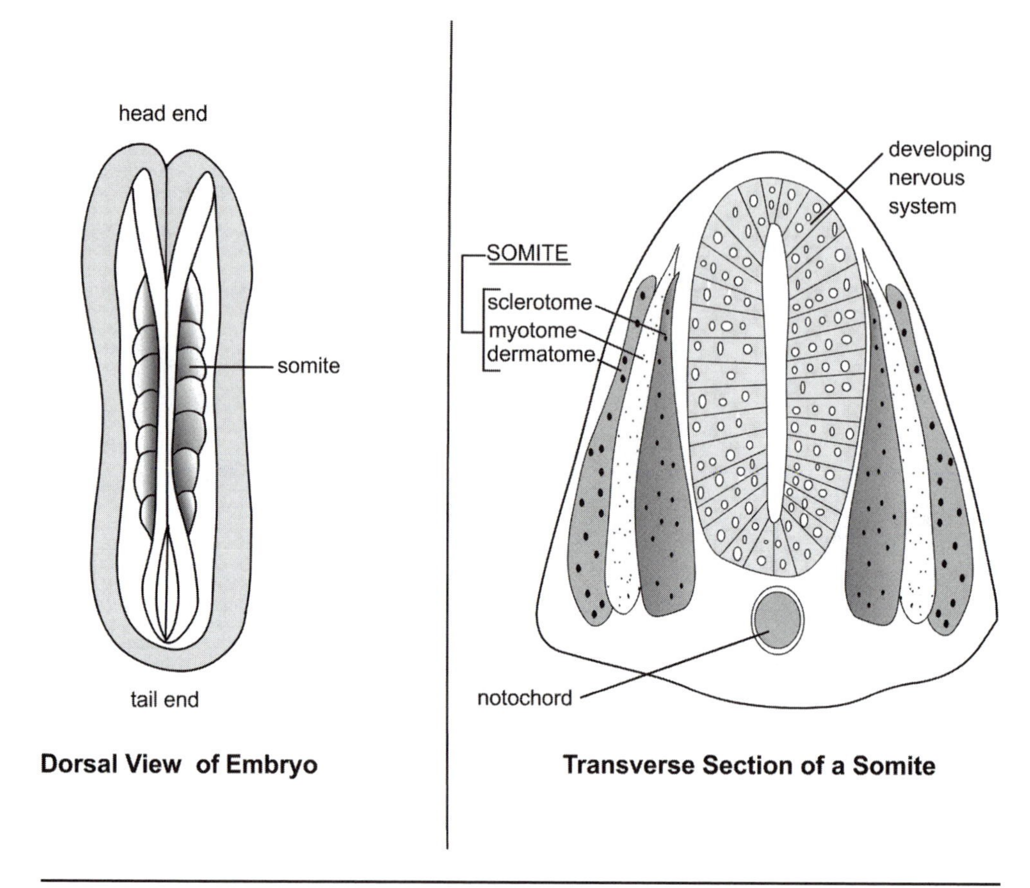

Figure 4.3. Location and structure of somites in the developing embryo.

Although cells within the somite all start out as identical mesoderm cells, they take on different fates depending on where they are located within the somite. Those cells in the front-facing region of the somite form what is called the scleratome and go on to form ribs, vertebrae, and cartilage. Cells in the rear portion of the somite form what is called the dermomyotome, which eventually goes on to form skeletal muscles. Figure 4.3 shows the somites in a developing embryo. The dermomyotome is further broken down into the region closest to the notochord, which produces cells that will form all the muscles of the back and neck (also called epaxial muscles), and the region farthest from the notochord and closest to the stomach, which produces all the muscles of the limbs and abdomen (also called the hypaxial muscles). Figure 4.4 shows the location of the epaxial and hypaxial muscles.

Cells in the somite need clues from the surrounding tissues in order to compartmentalize into these distinct regions. It turns out that the notochord

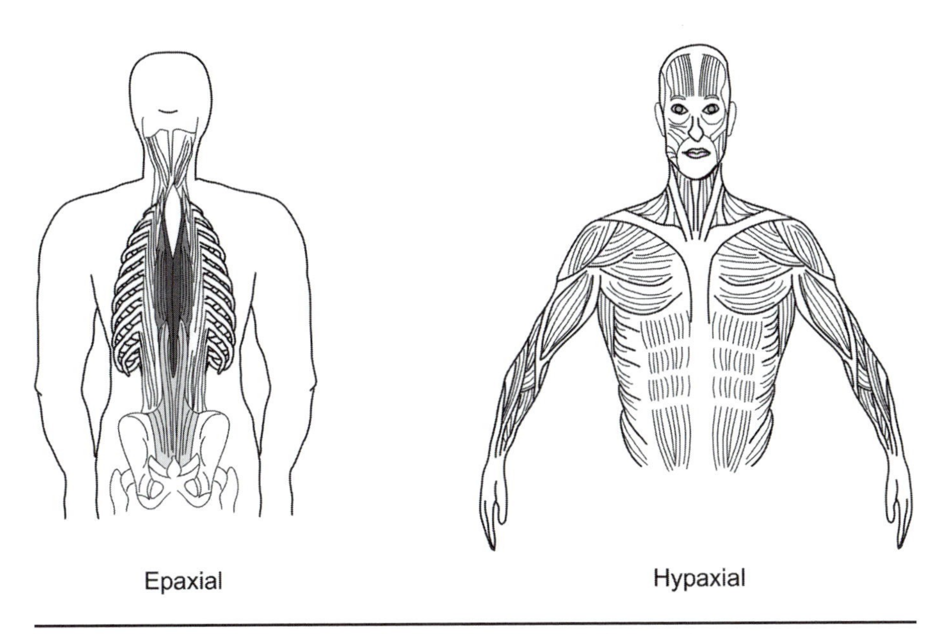

Figure 4.4. Location of the hypaxial and epaxial muscles.

releases a protein that informs neighboring muscle precursors in the somite that they are close to the notochord and should form the epaxial muscles. In response to this signal, those cells begin producing a muscle-specific protein called **MyoD**. From this point on, those cells produce proteins and respond to signals in a manner that is distinct from the hypaxial muscle precursors. Another signal from the forming neural tube directs cells to form muscles of the limb and abdomen.

When researchers prevent signals from either the notochord or neural tube from reaching the somite in laboratory animals, the muscle precursors are not able to develop into muscles. If only one of the signals is blocked, then the cells that did receive their developmental cue go on to develop normally while the other muscles fail to develop.

Migrating to the Limbs

While they are in the somite, the muscle precursor cells, called **myoblasts**, begin to make muscle-specific proteins, but they are not yet recognizable as muscles. They are also far from where muscles themselves will eventually form. The process of leaving the somite to form a muscle happens in two steps. First, a small subset of muscle precursor cells migrates from the somite to the location of a future muscle. This first wave of cells is called the **primary myoblasts** or embryonic muscle. These cells elongate and fuse into muscle cells that establish the location of the eventual muscle, spanning between two developing bones and forming the initial framework for the body's musculature. Although the myoblasts do fuse to become cells with more than one nuclei, considerably fewer cells fuse in primary muscle than in later muscles, and these muscle fibers end up with fewer nuclei per cell. Primary muscles make up very little of the muscle mass and instead define the muscle's position.

The primary fibers form a scaffolding onto which the next wave of fibers attaches to fill out the size of the muscle. The second wave of fibers is called the **secondary myoblasts** or fetal muscle. Researchers don't know if the primary and secondary fibers come from a different region of the somite or if cells leave the same region in two separate migrations. Either way, when these cells reach the primary muscle, they fuse to form elongated muscle fibers with multiple nuclei and join the growing muscle. The secondary cells first form clumps of short muscle fibers along the length of the primary fiber, but these eventually fuse and elongate into fibers that span the entire length of the muscle.

The primary cells generally become slow-twitch muscles, while the secondary cells, which make up the bulk of the muscle, become primarily fast-twitch type IIa or IIb muscles, depending on the type of muscle being formed and the person's genetic background. The two waves of muscles also make a different form of myosin when they are developing. The primary muscles make a form of myosin called embryonic myosin, while the secondary fibers produce fetal myosin. Both of these are a slow form of the protein, though they both switch over to making adult forms of myosin soon after birth.

Muscle-Specific Genes in Development

A myoblast in the somite plays a different role in the embryo than a myoblast that is migrating or in the process of fusing to become a muscle. In each of these phases, the cells need a different set of proteins to propel the cells one step further down the developmental pathway to become muscle cells. Which protein the myoblast contains is controlled by which genes are turned on in that cell.

The first nudge to begin developing as a muscle comes from either the notochord or neural tube, which cause the myoblast to begin making the initial muscle proteins. These first proteins play the critical role of activating the next wave of proteins that will be needed in the further phases of development. Likewise, with each new developmental phase, existing proteins activate genes that will be needed in a later developmental phase.

This orderly process can be altered by signals from surrounding tissues. Some neighboring tissues may make a protein that prevents the cell from maturing too quickly or that tells a developing myoblast to become one kind of muscle over another. Altogether, these protein signals from within and without direct the cell in its progress from somite to mature muscle.

MYOGENIC REGULATOR FACTORS

The first protein that controls genes involved in muscle development was discovered in the 1980s, when scientists did experiments fusing a myoblast and a different cell type in a test tube. These fused cells contained a muscle nucleus, a nucleus from the other cell, plus all the molecules that were present in both the original cells. Under these conditions, the nonmuscle nucleus began using genes that make muscle-specific proteins. This result suggested that some proteins from the muscle cell were able to enter the nonmuscle nucleus and activate muscle-specific genes. This protein was eventually discovered and named MyoD. Since that time, scientists have found several proteins called **myogenic regulator factors (MRF)** that control genes throughout muscle cell development and that maintain the muscle cell once it is formed.

Much of what researchers know about genes that are used in muscle cells comes from studying myoblasts in a lab dish—also called in vitro experiments ("in vitro" means "in glass"). When the researchers expose the cells to different compounds or MRFs, they can then watch how the muscle cells develop. They can also examine what proteins are made by cells in different stages of differentiating in a lab dish. The researchers can then piece together how different chemical signals guide the muscle cells to develop within the embryo.

In one such experiment, researchers found that when they put a small amount of developing notochord near myoblasts, those cells started making the first MRFs. The researchers were able to isolate the protein made by the notochord that caused this change in the myoblast. This protein is what the notochord releases in the embryo to send neighboring somite cells down the pathway to become epaxial muscles.

Even at this early stage, the MRFs are extremely specific—the hypaxial myoblasts make different proteins than the epaxial myoblasts in response to the two signals that start these cells down their developmental paths. Later, as the cells migrate to the limbs, different MRFs direct the cells to make proteins involved in migration, fusing to form myotubes, and producing the sarcomere proteins.

CADHERINS

One protein, called a cadherin, is a good example of the kinds of protein changes that take place in a muscle precursor cell as it migrates and matures into a muscle fiber. Cadherin is used by cells to connect to other cells.

Migrating muscle precursor cells don't make cadherins, whereas the cells that have reached the muscle location receive a signal that causes them to start making cadherins. As the cells begin fusing, they only have cadherins in clusters where cells touch. Then, as the cells mature into normal muscle fibers they only produce a small amount of cadherin. In mature muscle, the cadherin is located where a muscle cell is in contact with a satellite cell. When a muscle is injured and requires satellite cells to fuse and repair the damage, the muscle cells will produce more cadherin, much like an immature muscle cell. The satellite cell then fuses, and the muscle once again produces very little amounts of cadherin.

Regulating the Pace of Development

DEVELOPMENTAL SIGNALS

Both the notochord and neural tube make proteins that tell regions of the somite to begin developing as muscle cells. These molecules, called sonic hedgehog and Wnt are important developmental regulators throughout the embryo. But those cells need to be kept in check to prevent them from developing before they migrate to the appropriate location. Another protein made by nearby mesoderm cells, called BMP-4, holds the myoblasts in check and prevents them from become full-fledged muscles while still in the somite. If researchers artificially prevent the somites from receiving that signal, the premature hypaxial and epaxial muscle precursors develop into muscles prematurely rather than migrating. As the muscles migrate and move farther away from the somite—and from the signal from the surrounding mesoderm preventing them from maturing—they can continue the development that began in the somite.

One critical step in this development is when myoblasts lose the ability to divide. Early myoblasts in the somite and while migrating can divide to produce new muscle precursor cells. In this way, the small number of cells in the somite produces a much larger number of mature muscle cells. While migrating, the myoblasts make a protein called **pax-3**, which helps maintain the cells in a state where they can divide. When they reach the limbs, the myoblasts begin making one of the most common muscle MRFs, MyoD, and lose the ability to divide.

In mice, researchers have experimentally prevented pax-3 from being made, and the muscles that should have divided many times while migrating to the limbs mature too early. Because those myoblasts lose the ability to divide earlier in their development, they also produce fewer eventual muscle cells. These mice had much smaller muscles than normal mice. In fact, mice that fail to make any pax-3 are born with no muscles in the limbs.

Another signal that controls myoblast division comes from the muscle itself. When myoblasts fuse, they begin making a protein called **transforming growth factor beta (TGF-ß)**. This molecule binds to proteins that dot the outside of neighboring myoblasts. Once bound, TGF-ß prevents those cells from fusing with the developing muscle. This developmentally arresting process is probably what causes some myoblasts to stay on the outside of the muscle as satellite cells. Myoblasts that have already begun to fuse and form mature muscle don't have the appropriate protein on their surface and can't recognize the signal to halt their development.

A related protein, called **fibroblast growth factor (FGF)**, also maintains myoblasts in a state where they continue dividing. In another mouse experiment, researchers found that muscles that didn't have FGF signaling had reduced muscle size. Without FGF to keep the cells in a proliferating state, the cells divided too few times and produced smaller muscles.

SATELLITE CELLS

Once the cells reach their final destination, they go through many rapid divisions to increase their numbers, then begin differentiating into muscle cells that fuse and fill out the growing muscle. Although most cells get the signal to mature, about 5 percent of the cells remain as undifferentiated satellite cells. These cells are still immature and retain the characteristics of those cells that have migrated to the muscle but have not yet fused. They also retain the ability to divide, which mature muscle cells lose when they fuse. These satellite cells divide in response to muscle injury, fusing with the damaged region to repair the muscle.

Satellite cells studded along the outside of muscle cells are in a state of arrested development until they are called upon to repair damaged tissue. They multiply many times, leaving one progeny to remain as a satellite cell while the rest mature, lose the ability to divide, and fuse with existing muscle cells.

It appears that the processes used in development also come into play when satellite cells repair muscles. One signal from damaged muscle is FGF—the embryonic signal that maintains myoblasts in a dividing state. This signal also encourages nearby satellite cells to begin dividing after muscle injury.

Although satellite cells have begun developing into muscle cells and produce some muscle-specific proteins, they don't make MyoD. When a satellite cell becomes activated by damaged tissue, it begins making both MyoD and other MRFs that are involved in muscle development. As the cells mature, they produce the muscle-specific proteins in the same order as is seen in early muscle precursor cells as they fuse to become part of a larger muscle.

Controlling Muscle Size

The hand contains small, delicate muscles that control intricate movements. The hamstring and buttocks contain large, powerful muscles. Yet all the muscles of the body come from somites that start out about the same size. The eventual size of a muscle comes not from how many cells migrate to that muscle from the somite but from how many times those cells divide. A small number of cells might migrate to the hand, divide only a few times, then fuse to form the various hand muscles. The migratory cells that form the hamstrings must divide many more times in order to generate enough cells to make up the large hamstring muscle. Although in some cases more fibers might be added to the muscle in adulthood, muscles generally grow larger by the expansion of muscle fibers formed during fetal development. For this reason, the number of fibers made during development is critical for normal adult muscles.

The signals that tell cells to keep dividing come from within the migrating cells themselves and from the tissues surrounding the location of the eventual muscle. If that division process halts early—or is halted artificially in the lab—the initial size of the muscle is smaller, and it remains smaller throughout the animal's life.

Determining how long to allow myoblasts to continue dividing is a delicate balancing act for a developing organism. If division halts too early, the muscle remains small. If division goes on too long, the muscles become excessively large. Another risk to ongoing division is that the cells may lose control of their division and become cancerous. This is where genes such as pax-3, TGF-β, and FGF come in to play. They ensure that the myoblasts are sufficiently mature before developing, but also prevent the myoblasts from maturing into muscle cells as soon as they reach the limbs.

Throughout life, a protein called **insulin-like growth factor (IGF)** seems to maintain the size of the healthy muscle. Muscles that receive additional doses of IGF grow in size and also grow stronger. These muscles also have a slight shift toward fast-twitch fibers. The added IGF has the additional advantage of making the muscle more able to regenerate after damage. One possible role for IGF in a normal muscle is inducing satellite cells to mature and fuse with existing muscle fibers, leading to increased muscle size and strength gains.

Connecting to Nerves

Throughout the time that the myoblasts are fusing to form muscle cells, nerves are also migrating from the spinal column and forming connections with the muscles. The nerves in the spinal column send out projections that follow the trail of primary myoblasts as they migrate. The growing tip of this nerve projection, called the growth cone, pulls the projection along as the nerve expands. This growth cone stops growing when it encounters a muscle cell that is in a developmental state in which it is ready to form nerve connections. At this point, the nerve forms a structure called a synapse where the signal to contract is transferred from the nerve to the muscle. Within about eighteen hours of coming in contact with the muscle, the nerve side of the synapse contains the chemical that is released to activate the muscle. This chemical, called acetylcholine, is located with small capsules in the nerve side of the synapse.

FORMING NERVE CONNECTIONS

In the adult muscle, each muscle fiber has only one connection with a nerve, but as the nerves and muscles are developing they form many preliminary connections with the primary muscle. During this time, the primary muscle has receptors for acetylcholine all along the membrane. Eventually only one nerve maintains a connection with each muscle cell, and the acetylcholine receptors become localized to where the synapse with that nerve forms. This single nerve connection is usually located near the center of the muscle fiber.

The nerve connections to primary fibers act as a signal to the secondary myoblasts that are migrating to the developing muscle. These signals stimulate the secondary myoblasts to divide and produce more cells that will eventually fuse to form the secondary muscle fibers. These myoblasts generally attach to primary muscle at sites that are close to where nerves have formed connections. Once the secondary fibers have formed, they make a connection with only one nerve.

The initial junction between the nerve and muscle grows stronger in response to signals from both the nerve and the muscle. Initially, when many nerves attach to a primary fiber, those nerves all begin firing. The nerve connections that don't generate much response from the muscle deteriorate while those that generate a strong response continue developing into a mature nerve connection. Researchers think that the muscle releases a substance that is sent to the body of the neuron, signaling that neuron to strengthen the nerve-muscle junction. In response to that stronger connection, the muscle cell begins producing additional muscle-specific genes that help the muscle mature.

This process of signaling a muscle by maturing nerves can happen singly but often occurs in antagonistic pairs. For example, nerves to the biceps and triceps muscles might fire at the same time. This activity accounts for the stretching motions that can sometimes be seen in an ultrasound of a developing fetus. The act of contracting both the biceps and triceps causes the arm to stretch out, just as contracting both the hamstring and quadriceps causes the leg to extend. Although the stretching makes a fetus look as if it is resting calmly, it is actually a sign that the muscle development is progressing normally.

NERVE CONNECTIONS REGULATE MUSCLE DEVELOPMENT

In both primary and secondary muscle fibers, not having a proper nerve signal can prevent normal development. In animals, researchers have prevented the nerve connection from forming with developing muscle fibers. When they do this to primary fibers, those fibers still form the initial muscle, but not all fibers that are formed survive. The secondary fibers will also form without a nerve connection, but they form smaller muscles with fewer eventual fibers. Although muscles that develop in the absence of a nerve signal are smaller, they do contain a diverse range of fiber types, including slow-twitch fibers and both forms of fast-twitch fibers. Later in life, the type of nerve connection a muscle has can determine whether that fiber is a slow- or fast-twitch fiber. It appears that the muscles initially develop the full range of different fiber types without input from nerves.

A healthy nerve connection is important in maintaining adult muscles as well as in developing muscles. If a muscle no longer receives a nerve signal due to an injury, that muscle withers away even if the muscle receives an artificial electrical signal to maintain some muscle tone. In animals, when researchers remove a muscle that has lost its nerve connection and test its strength in a lab, they find that the muscle can't contract normally in response to an electrical signal. It appears that without the proper nerve connection, the muscle lacks a signal to maintain itself.

At the same time that nerves that cause the muscle to contract—also called motor nerves—are establishing connections, their sensory counterparts are forming connections with the developing muscle. These nerves have their point of contact with the muscle encapsulated in fibrous tissue, and they also form connections within the tendons that hold the muscle to the bones. These nerves sense when the muscle or tendon is stretched either by carrying a heavy load or by simple stretching.

Unlike motor nerves, removing a sensory nerve from a developing muscle does not alter the course of development. However, muscles without sensory nerves may not respond normally to exercise in an adult, because the muscle cannot sense stretch. Without the ability to sense when the muscle has stretched too far, muscles lacking sensory nerves may also be more susceptible to injury.

Determining Myosin Type

The first muscles to form make a type of myosin called embryonic myosin. This protein uses ATP slowly, much like the myosins found in slow-twitch muscles. The second wave of muscle fibers also make a slow form of myosin, called fetal myosin. Although both primary and secondary fibers start out making a slow myosin, the primary fibers generally go on to become slow-twitch fibers whereas the secondary fibers can become either slow- or fast-twitch fibers in the adult.

The muscles continue to use these early forms of myosin until soon after birth. At this time, muscles stop making the fetal and embryonic forms of myosin and begin making one of the many adult myosin proteins. Which myosin type the fiber makes determines whether that fiber will go on to become a slow-twitch or fast-twitch fiber. Although this switch to adult myosin types is to some degree regulated—calf muscles, for example, contain a large number of slow-twitch fibers in all people—

the eventual distribution of fast-twitch and slow-twitch fibers is unique in each person.

Several signals control what type of myosin a fiber should make. All fibers in a motor group connect to a single nerve and are the same fiber type. Experiments have shown that connecting a fast-twitch nerve to a slow-twitch fiber will at least partially switch that fiber's type, and vice versa. What appears to determine fiber type, at least in part, is speed of the electrical signal coming from the nerve.

When a nerve stimulates a muscle, calcium is released from the sarcoplasmic reticulum. Slow-twitch muscle fibers tend to have a slow, continuous signal from the nerve, which doesn't give the cell enough time to sequester calcium in the sarcoplasmic reticulum between signals. For this reason, slow-twitch fibers tend to contain more calcium than fast-twitch fibers. When researchers artificially stimulate fast-twitch fibers continuously, raising calcium levels, those fast-twitch fibers take on some characteristics of slow-twitch fibers. Likewise, when the researchers artificially lowered calcium levels in the slow-twitch fibers, those fibers produced proteins found in fast-twitch fibers. From these experiments, some researchers think that it is the amount of calcium in the fiber that determines what type of fiber it will become, at least in the adult.

MUSCLE GROWTH IN CHILDREN

Muscle development doesn't end at birth. All muscles continue to grow and change throughout childhood and puberty. This growth is evident by muscle weight in infants and adults; at birth, muscles make up about 25 percent of a baby's weight. That increases to 50 percent or more of an adult's weight.

The most dramatic increase in muscle mass comes during puberty, particularly in boys. When they enter puberty, testosterone shoots up to levels ten times higher than before puberty. Among its many roles in the body, testosterone causes muscles cells to increase in size—the muscle cells literally swell with added fibers in response to the hormone. The muscles also increase in length to accommodate the lengthening bones.

In women, muscles make up roughly 40 percent of adult weight. As with boys, most of the increase comes during puberty when hormones related to testosterone cause muscles to grow, in addition to other physical changes. Girls reach their peak muscle mass at around ages 16 to 20, while boys peak at about 18 to 25.

The decreasing amount of muscle after the late teens or early twenties can be both biological and social. Girls tend to take up more sedentary activities after high school. Without regular activity to maintain the muscles, they tend to become smaller. Likewise, boys often become less active after college and have a similar decrease in muscle size. When people keep exercising throughout their life, they can hang on to their peak muscle mass well into adulthood.

In addition to changes in muscle size, the muscles become more coordinated with age. This change in coordination has more to do with the nerves going to the muscles than with the muscles themselves. Nerves have a coating called a myelin sheath (see the Nervous System volume of this series for more details about the anatomy of a nerve). This sheath is a fatty protective layer that speeds the signal along the nerve. Without the myelin sheath, the signal travels slowly from the brain to the muscle. When this myelin sheath has completed development on all nerves, which usually happens around puberty, the muscles receive the signals at the fastest possible speed and can react quickly to decisions to move.

MUSCLES IN AGING PEOPLE

Changes to muscles don't end at puberty. Throughout life the muscles respond to exercise or injury by increasing or decreasing in size, by changing fiber type, or by modifying enzymes made in the fibers. As people get older than around age 50, additional changes start taking place. The muscles become weaker and less able to respond to training. People who continue to exercise regularly can make up for some of this loss of strength, but even with training the muscles are never as strong as when a person was younger.

The most obvious difference in older people's muscles is the smaller size. The muscles have fewer signals such as testosterone or IGF to maintain the muscles, so the fibers grow smaller and the muscles are surrounded by more fat than in younger people. The reduced strength may start with the nerves rather than the muscles. The rate at which nerves can send signals slows down with age. With fewer signals, the muscle doesn't contract as hard. The muscle also becomes smaller because it is not being called upon to lift as much weight. (See "IGF Maintains Weakening Muscles.")

Whether nerves or other factors underlie the weakening muscles, people over the age of 50 lose about 10 percent of their muscle fibers per decade. Those most likely to be lost are the fast-twitch fibers. In people who continue training, the muscle fiber distribution stays about the same as in younger people. Those who don't train end up with a greater proportion of slow-twitch to fast-twitch fibers than when they were younger because the fast-twitch fibers degenerated. Not only are the muscles getting smaller, but the fibers that remain aren't those that are best suited for heavy lifting.

In addition to losing muscle mass, athletes show about 1 percent decline in VO_2max per year after age 50. People who don't continue training lose even more conditioning, at a rate of about 15 percent of VO_2max per year.

IGF Maintains Weakening Muscles

One of the proteins that helps maintain muscles throughout life also seems to help prevent muscles from decreasing with age. People are starting to use this protein—called IGF—to keep older people strong and less likely to fall and injure themselves. Having more muscle mass also helps protect joints, bones, and vertebrae from getting injured in people who do fall. In some studies, older adults who received injections of IGF had muscles that were similar to normal muscles in young people in both size and strength. In particular, the IGF injection prevented the loss of fast-twitch fibers that is normally seen as people age. Doctors think that by giving people more IGF, the satellite cells can become activated to repair damage.

If this research with IGF continues to be successful, other uses may be down the road. In people who wear a cast to immobilize a broken bone, the muscles around that bone grow weak from lack of use. IGF injections could maintain those muscles while the bone heals. IGF may also be able to prevent the muscle wasting that takes place after many months in space. With no gravity to work against, muscles don't need to be very strong in space. Because they aren't getting heavy use, the muscles can atrophy so much that astronauts have a hard time walking when they get back to Earth. Some researchers see a future when IGF could help those astronauts hang on to their strength even while on long trips in low gravity.

Researchers aren't sure if the decrease in VO_2max comes from the heart's inability to beat as hard or from the weakening muscles.

Some of the effects of older age can be helped by regular exercise. Endurance-type activities such as running help muscles maintain their ability to use oxygen, but don't seem to help keep muscle mass. Strength training, such as weight routines at the gym, has the opposite effect. It helps keep the muscles strong but doesn't help the ability to use oxygen, which is similar to strength training in younger people.

HEART MUSCLE DEVELOPMENT

The heart is one of the first organs to begin developing in an embryo, and it is also one of the most complex. Ten percent of spontaneous abortions are caused by heart malformations, and heart defects make up more than a third of all congenital defects. Of these defects, the most common are problems with how the valves between the heart chambers function and with an underdeveloped left ventricle—the heart chamber that pumps blood to the body. These most common problems occur in eight out of every 1,000 babies born.

The heart muscle comes primarily from mesoderm cells in the front portion of the developing embryo. These migrate to the region that will eventually contain the heart. Once there, signals from the overlying ectoderm—the cells that will eventually form skin and nerves—direct the muscle precursors to begin forming the heart.

The first step in heart development occurs when the myoblasts form a tube that then loops around to the left to form twin, parallel tubes. These make up the left and right sides of the heart. The upper region of these loops becomes the atria and the lower region becomes the ventricles. Soon after the tube loops around, the atria and ventricles begin making their own, unique subset of heart-specific genes.

At about twenty-one days after fertilization, the cells of the primitive paired tubes begin beating. This heartbeat is one of the most visible traits on a pregnant woman's first ultrasound. The initial beating cells aren't in the form of proper muscle. Rather, they are individual cells that contain actin and myosin in a somewhat organized arrangement, and they can contract.

After the cells start beating, they elongate and fuse into cardiac muscles. The actin and myosin filaments align so the beating of the individual cells contributes to a squeezing motion in a single direction. The cells also develop intricate interconnections, so a single nerve signal can cause the entire heart muscle to contract in a regular heartbeat. While the heart is developing the cells continue to divide, increasing the size of the heart. It isn't until around birth when the heart muscles lose the ability to divide. Throughout childhood, the heart grows in size by the existing muscle fibers expanding rather than by new fibers being added.

In addition to muscle, the heart has arteries that provide blood to the beating heart. These comes from cells in the neural region of the embryo. They migrate to the outer layer of the heart where they form arteries, including the large arteries leaving the heart and leading out to the body. (See "Differences between the Left and Right Sides of the Body.")

SMOOTH MUSCLE DEVELOPMENT

Smooth muscle is much more variable than either skeletal or cardiac muscle. It forms tubes around arteries, sheets to make up the bladder, and the tiny muscles that control the size of the pupil in response to light. With so many roles in so many locations throughout the body, researchers haven't pinned down one developmental pathway that the cells follow. The task of studying smooth muscle development is made harder by the fact that smooth muscles change into other cell types in a lab dish. Unlike skeletal and cardiac muscle, which scientists can manipulate in the lab to learn how factors affect their development, smooth muscles can change developmental course to become skeletal muscle or to form diverse kinds of smooth muscle.

Differences between the Left and Right Sides of the Body

In every person's body, the left and right sides are slightly different. This is true not only in what hand a person prefers to use, but also internally. The liver has more lobes on one side than the other, the spleen is located on one side of the body, and the heart is tilted toward the left side.

The primitive heart tube looping around is the first sign in embryonic development that the body is asymmetrical. Until that time, the embryo has obvious differences between the back and the front, and between the head and the eventual feet, but left and right are identical. Researchers are only now beginning to understand the signals that tell the heart tube to loop one way rather than the other.

In some people, the signal to loop the appropriate direction is misdirected, and the heart loops the other direction. These people have a heartbeat closer to the right side of their chest and have other organs reversed. This switch doesn't cause any biological problems, but it can be confusing for an examining doctor.

Another problem with studying smooth muscle is finding them in the embryo. Both skeletal and cardiac muscle make specific proteins that scientists can look for to identify those cell types in mice or other lab animals. Smooth muscles make forms of myosin that are sometimes found in the other muscle types, making it hard to pick out developing smooth muscle cells.

Despite these difficulties, researchers have learned that smooth muscle comes from both the mesoderm cells and the ectoderm cells that make up part of the nervous system. In particular, the smooth muscles that make up the arteries leaving the heart come from tissue that also forms nerves. The various smooth muscle tissues develop at different times throughout the body. Developing babies can swallow at ten weeks, and at thirty-four weeks they can push material through the intestinal system. Some muscles of the blood vessels don't become fully functional until after birth. These changes may take place because of change in the blood supply after the baby is born.

As early smooth muscle cells begin forming tissues, the cells elongate into a dense layer of parallel cells. These make actin and myosin filaments along the long axis of the cells, so when the tissue contracts all cells contract in the same direction. In some tissues, these smooth muscle cells keep dividing until birth—unlike skeletal muscles, which lose the ability to divide when they fuse to form the muscles. The number of smooth muscle cells increases by three to five times from the time they are formed. The cells themselves also grow larger to produce the overall gains in muscle size.

Smooth muscles are less dependant on nerves for development than are skeletal muscles. When researchers prevent nerves from reaching smooth muscles in lab animals, the muscles still develop normally. In some organs, such as the bladder, the adult smooth muscles rely completely on signals from nerves in order to contract. If the nerve to the bladder becomes damaged, the smooth muscles are paralyzed. In other tissues, the smooth muscles contract in response to signals between cells, stretch receptors, or hormones for the cue to contract.

As with skeletal and heart muscles, smooth muscles can grow stronger in response to heavy use. This increase in size and strength is usually the result of muscle cells growing larger rather than dividing to create new cells, although these cells are also capable of dividing if growing larger doesn't bring enough strength gains.

Smooth muscles usually grow stronger in response to diseases or abnormal conditions rather than exercise or other lifestyle changes. If the exit from the bladder becomes partially blocked, for example, the bladder muscles become stronger to push the urine through the reduced opening. Furthermore, in the bladder some surrounding connective tissue can turn into smooth muscle tissue if the needs are great enough. Likewise, if the intestines become blocked, the muscles that push food through the obstruction grow larger. Overall, the adult smooth muscles are capable of growing to as much as ten times their original volume.

5

Early Discoveries in Muscle Anatomy and Physiology

Long before researchers understood how actin and myosin filaments slide across each other in a contracting muscle, ancient doctors and philosophers pondered how the body worked. In fact, the word muscle comes from the Greek *musculus*, or small mouse, in reference to the size and shape of some muscles. Greek theories about the function of muscles were extremely long-lasting, extending well into the 1700s. Theories began changing when physicians in the late 1700s discovered how nerves transmit impulses—discoveries that revolutionized thinking about how muscles contract. Since that time, the study of muscles advanced quickly, aided by research into nerves and metabolism that also hinted at how muscles might function.

Although researchers today have a good molecular understanding of how muscles function at rest and during exercise, many questions remain unanswered. The history of muscle research is still ongoing in the fields of physiology, cell biology, and exercise physiology, where physiologists of today are continuing to learn about muscle biology.

SKELETAL MUSCLE

Greek Theories about Skeletal Muscle Function

The Greeks developed models about how the body functioned that would remain the dominant theory until the Middle Ages. One of the earliest and most influential Greek physicians and philosophers was Aristotle (384–322 BCE). Although neither Aristotle nor those who immediately followed him formed any theories about how muscles contract, he did describe the physical geometry of locomotion, along with mathematical models for movement. Working at about the same time was Hippocrates (460–377 BCE), whose oath doctors still say before earning their medical degrees. Hippocrates was the first to publish a detailed anatomy book, based on dissections made on deceased criminals in Athens and Alexandria. These early physicians also studied muscular diseases such as forms of rigidity or convulsions and understood that these diseases were related to the muscles.

In around the second century CE, as Greece declined and the Roman Empire grew, most scientific progress was taking place in Rome. The most influential scientist working during this time was a physician and philosopher named Claudius Galen (131–201), who described muscle structure and function. Galen has been called "the father of experimental physiology" because of his important contributions to physiology research. According to Galen and other physicians of that time, all matter was made up of the four elements: fire, water, air, and earth. Matter was also thought to have qualities of being hot, wet, cold, and dry. People maintained their health by having the proper balance of these elements and qualities. The body was assumed to be kept alive by three animating forces, which resided in the brain, heart, and liver. The heart, in this scheme, was the *spiritus vitalis*, which was responsible for the pulse, circulation, and distribution of heat. The brain contained the *spiritus animalis*, which was the thinking soul and was responsible for the body's movement. The liver contained the *spiritus naturalis*, which was in charge of nutrition, blood formation, and metabolism.

Galen was one of the first to examine muscles in any detail, and he was the first to understand that muscles could contract independently from each other. Because dissection of humans was frowned upon in his day, Galen drew most of his conclusions from dissections of dogs, pigs, and apes. He considered only the skeletal muscles to be true muscles, while the heart, stomach, bladder, and uterus were "muscle-like" and could move autonomously. The skeletal muscles had two components in Galen's analysis—tendons and a fleshy, insensitive substance. He thought that muscles had an inherent ability to contract. This went against the prevailing thought, which held that muscle contraction was triggered by the animating force in the brain, which was transmitted to the muscles by nerves. Galen taught that the muscles contracted by swelling, which pulled the two ends closer together, and that the force of the contraction came from the tendons rather than the muscle itself. Despite his primitive understanding of how the body functioned, Galen gave a good, if simplistic, description of muscles. They do have the ability to contract and are triggered to do so by nerves.

Galen broke movement down into four types: (1) movement from a muscle contracting; (2) lengthening of the muscle when an antagonist contracts; (3) passive movement such as when an arm drops; and (4) lack of movement, such as when a person holds his or her arm at shoulder height. This last form of movement is actually a form of muscle-induced lack of movement, which Galen called "tonus," a precursor to how we use the words "muscle tone" today.

Galen was the first to recognize that muscles could contract even when removed from the body. Galen attributed this phenomenon to the muscle's innate ability to contract. From early experiments, he felt that signals from nerves instruct muscles to go against their innate ability and instead relax. But in experiments where he severed the nerve going to the muscle, the muscle lengthened and remain that way indefinitely. From this, he concluded that muscles require what he called "voluntary faculty from the soul," which arrived at muscles through the nerves, in order to contract. His book, *De Motu Musculorum* (On the Movements of Muscles), laid the foundation for muscle research and remained the leading book of its kind for the next 1,300 years.

The Renaissance Era of Muscle Research

Muscle research—and in fact all medical research—came to a standstill during the Middle Ages, which continued from about the fifth century to the fifteenth century. During this time, people turned to faith for an understanding of how things work rather than to observation and science. Research of any kind did not reemerge until the Renaissance began in the 1500s. During this period, two people who are primarily known as artists dramatically advanced the study of human muscles through careful drawings of cadavers.

One of these artists was Andrea Vesalius (1514–1564), who by age 28 had dissected all parts of the human body and published a volume of his drawings. These drawings were among the first to show, in great detail, the location of all the muscles of the body. In his book, Vesalius also describes experiments with living animals. In these, he observed that it was the fleshy parts of the muscle and not the tendons that attach the muscle to the bone that carry out the work of contracting, contrary to what Galen had believed. Vesalius also demonstrated that if a nerve is severed, the muscle can no longer contract. If the nerve is just injured, however, the muscle slowly regains the ability to contract as the nerve heals.

The other great artist of the 1500s who contributed to the study of muscles was Leonardo da Vinci (1452–1519), who did most of his work in the early 1500s. Leonardo was the ultimate Renaissance man, who excelled as a sculpture, painter, musician, poet, and scientist. His detailed sketches of human anatomy came from cadaver dissections that he did in collaboration

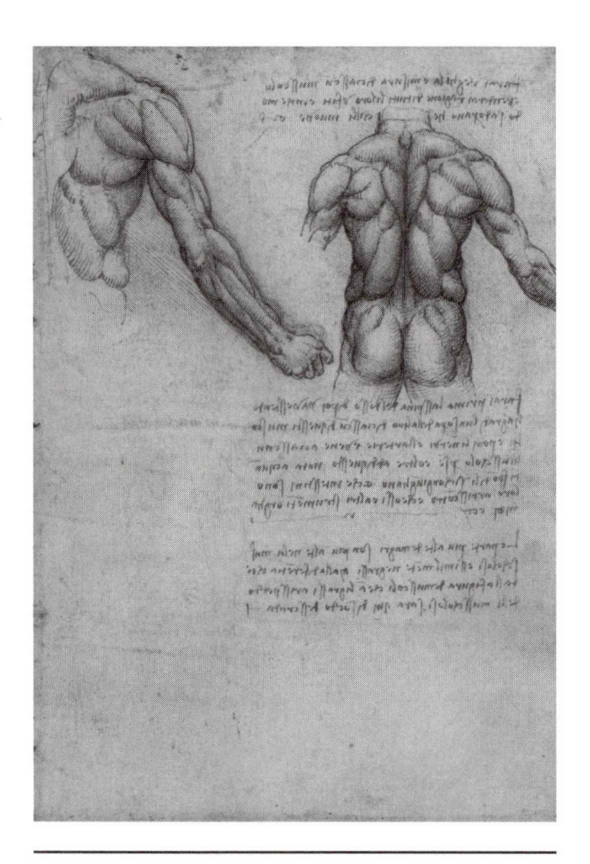

Study for human musculature, Leonardo da Vinci. © Scala/Art Resource, NY.

with the anatomist Marc Antonio della Torre (?–1511), with whom Leonardo intended to publish an anatomy text. Della Torre died before they could finish the project and the book never got published, but Leonardo's drawings survived, along with his notes about how he thought the organs functioned.

Among Leonardo's drawings are finely detailed sketches of the human muscular system. He is said to be the first to accurately draw all muscles of the human body, and his sketches are so complete they still adorn medical texts to this day. According to Leonardo, the muscles were made up of fine fibers connected to a nerve. The hollow center of the nerve carried a signal from the brain instructing the muscle to contract. Although some at this time thought that the muscle expanded with air, Leonardo believed that the muscle attracted the two tendons together through a signal that went along the fibers between the two tendons. In addition to studying the structure of the muscles themselves, Leonardo understood that opposing muscles could act synergistically.

Although Leonardo pioneered the thinking that the body worked through physical forces rather than animating spirits, his theories did not influence much of the thinking at the time. His notebooks were lost and therefore unavailable to later scholars, and they only reemerged in the 1900s, when an understanding of the muscular system had already gone far beyond what Leonardo had theorized. Without his contributions, physicians and scientists continued for some time to believe that animal spirits were required to influence muscles to contract.

The Classic Age of Scientific Discovery

The sixteenth and seventeenth centuries saw the rise of physics and chemistry and of physical descriptions for natural phenomena. This interest in descriptive science included advances in the understanding of how muscles contract. Among the first contributors to this modern age of muscle research was Fabricius of Aquapendente (1537–1619), who first analyzed the geometry of muscle movement. He was quickly followed by a flurry of research into muscle physiology that took place across Europe. Among these physiologists, the first to publish and therefore communicate this new way of thinking to the rest of the world was the Englishman William Croone (1633–1684), who worked in the mid-seventeenth century. Croone still believed in the animal spirits that drove muscles to contract, but he did understand that it was the fleshy parts of the muscle that did the contracting. He was also the first to describe the importance of a functioning circulatory system in muscle physiology.

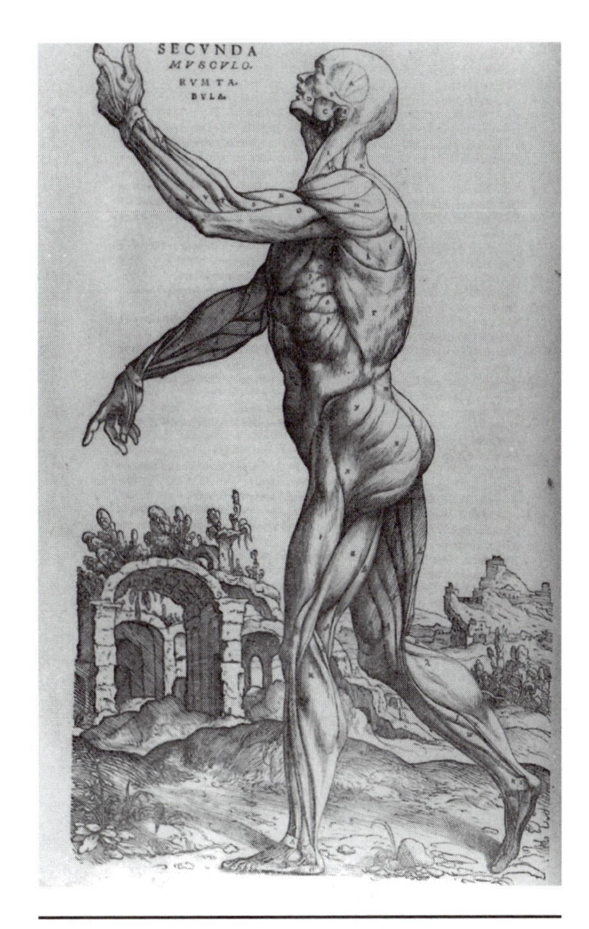

Human muscular drawing by Andreas Vesalius. © National Library of Medicine.

In addition to his detailed drawings of how nerves and blood vessels interact with muscles, Croone did geometric analysis of how muscles contract across a joint to bring two bones closer together. In his will, Croone asked to have an annual lecture on the physiology of what he called "muscular motion" held in his name. This lecture series, now called the Croonian lectures, still continues in England to this day.

A contemporary and friend of Croone's, Nicolas Steno (1638–1686) (who is often referenced as Nicholas Stenson), had access to a microscope, which had just been invented in 1660, and with that laid out the foundation of muscular function. Steno understood that the fibers he could see under the microscope were the workhorses of the muscle. He felt that it was the sum of these fibers contracting individually that led to an overall muscle contraction. His work with the mathematics of muscle contraction is often considered to be the beginning of the field of biomechanics.

The inventor of the microscope—the prolific naturalist Anton van Leeuwenhoek (1632–1723)—was the first to describe seeing the stripes in skeletal muscle that give the tissue the name striated muscle. He inaccurately characterized these stripes as being due to an additional fiber coiling around the central long strand. Other researchers described seeing stripes in skeletal muscles but not in the muscles that line the bladder and uterus.

This was the first time the two muscles were broken down into the categories of "striated" and "smooth" muscles. Looking at thin slices from a muscle cross-section, James Keill (1673–1719) used the microscope to investigate how many muscle fibers a muscle contained. By dividing the maximum amount of weight a muscle could lift before dissecting the muscle, Keill also worked out how much weight per fiber a muscle could lift. This calculation can never be entirely accurate, because each fiber may have a different number of myofibrils. It is the width of a fiber (and therefore the number of myofibrils it contains) that determines the fiber's strength.

Steno had no solid explanation for how muscles contract, but he did not follow the idea that animal spirits flowed down the nerves to animate the muscles. His misgivings were shared by his contemporary Francis Glisson (1597–1677; see photo), who finally put to rest the idea that a fluid poured into the muscle during a contraction. Glisson reasoned that if a substance entered the muscle while contracting, the muscle volume would go up. He developed a device in which a man's muscle could be completely submersed in water, then had the man extend and contract his muscle. If the muscle increased in volume during the contraction, the water would be displaced and move higher in the apparatus. Through this, he found that, if anything, the total volume was slightly smaller when the muscle contracted. Although Glisson's work was published first, the Dutch naturalist Jan Swammerdam (1637–1680) probably did his experiments with the same conclusion twenty years before Glisson, though his book was not published until many years later. The photo shows the apparatus Swammerdam used in his experiments.

Francis Glisson teaching an anatomy class. © National Library of Medicine.

In addition to studying the geometry of movement and the nature of what causes muscles to contract, experiments were drawing people closer to understanding the mechanics of how the fibers grew shorter in a contraction. In particular, Giovanni Alphonso Borelli (1608–1679) understood that tonic contraction—constant contraction that supports the joints without moving them—is necessary to prevent antagonistic muscles from overextending. He believed that the force behind muscle fibers contracting came from a fine, lattice-like mesh that covered the inner walls of the muscle fibers. The squares of this mesh could extend lengthwise into a long, flat rhomboid shape in a relaxed muscle, then convert into a shortened, taller rhomboid when the muscle contracts. This action is very much like a lattice gate that can expand and contract across a doorway to keep a child or animal from entering a room or stairway.

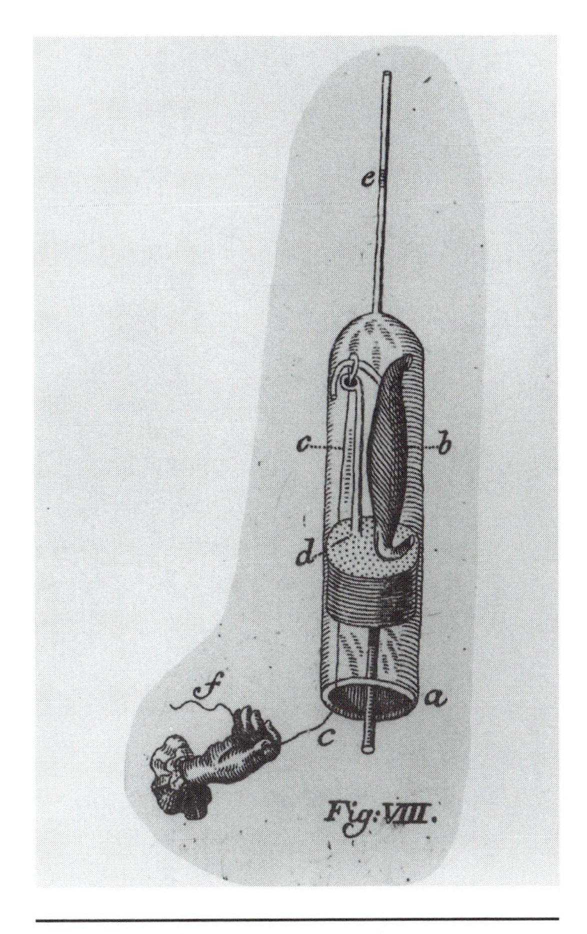

Apparatus used by Swammerdam for measuring contraction in a frog muscle. © National Library of Medicine.

The Enlightenment Era of Muscle Research

In the early eighteenth century, a new era began that was later dubbed the Enlightenment. During this time, philosophers wrote that knowledge had to be based on experiences. Gone were the days when faith or mysterious animating spirits could account for physical phenomenon. It was during this period when German philosopher Immanuel Kant (1724–1804) wrote that the motto of the Enlightenment was "Have the courage to use your own intelligence." In science, the Enlightenment produced steam engines, air balloons, manufacturing, weaving mills, and other forms of practical industry.

This very practical era brought Herman Boerhaave (1668–1738), who wrote that the human body was a machine whose solid parts acted like mechanical instruments (see photo). Muscle fibers were simply hollow extensions of the nerves surrounded by a lubricating fluid. The muscle contracted when fluid from the nervous system entered the muscle fiber.

One of Boerhaave's students, Albrecht von Haller (1708–1777), wrote many volumes on various aspects of physiology, but his passion seemed to be a theory about the intrinsic properties of some tissues, including muscles. Von Haller graduated at age 15 and began practicing medicine at 21. Not satisfied with medicine alone, he went on to become a professor at the

Hermann Boerhaave. © National Library of Medicine.

newly founded University of Göttingen at 28, where he founded an anatomical institute and participated in more than 360 dissections. Between his academic work and extensive publications, von Haller is said to have slept only four hours per day.

In von Haller's view, the tissues of the body were all made up of various types of fibers held together by a gelatinous substance. One group of fibers made up membranes, vessels, or the coverings on organs. Nerves, which made up the second type of fibers, contained the property of sensibility. They were the only type of fiber that could feel pain. The third type of fibers, which made up all muscles, could not feel pain but could respond to stimuli and contained the property of irritability. According to von Haller, it is the intrinsic irritability that causes muscles to contract in response to a signal from the nerves. He noted that because the heart and intestines can contract even in the absence of a nerve connection, those muscles must become irritated by some other stimuli, while skeletal muscles rely exclusively on nerves to trigger their irritability. He also discovered that although nerves do trigger a muscle's "irritability," the nerves could also fire in the absence of a muscle, making the nerves critical to, but independent from, the muscles. With his extensive publications, von Haller's ideas about sensibility and irritability had a profound influence on many biologists of his day.

One of von Haller's contemporaries, the leading English scientist Robert Whytt (1714–1766), also turned his attention to the problem of muscle contraction. Among his discoveries is the first clear description of a reflex reaction. He also carefully deconstructed arguments put forward by previous researchers about the nature of contraction. However, he never came up with an explanation of his own for how the muscles contracted. Some of Whytt's ideas brought him in conflict with von Haller, who believed that some muscles can contract in the absence of a nerve signal. Whytt held that some muscular contractions, such as reflexes, came from signals that a person might not even be aware of—signals other than the pain that von Haller focused his attention on. The discovery of a reflex reaction, in which a signal for a muscle to contract comes from nerves in the spinal column but not in the brain, is attributed to Whytt.

The Discovery of Electrical Currents in Muscles

Around 1780, the discovery that living tissues react to electricity forever changed the study of muscles. For nearly the next century, questions about the body's source of electricity, how that electricity is transferred to the muscles, and how the muscles respond to the signal would consume muscle research. This avenue of research was driven, in part, by the important discoveries being made with electricity in other fields—electricity was clearly the hot field of the day and remained that way until the late 1800s when scientists finally pinned down the nature of the body's response to electricity.

This initial discovery that bodies can respond to electricity came from a chance observation by Luigi Galvani (1737–1798) and his wife Lucia (1743–1790) that frog legs hung from copper hooks would spasm when in contact with the iron supporting structure. At this time, frogs and dissected frogs' legs were widely used for studying nerves and muscles. Previous researchers had reported that isolated frog legs could twitch and that decapitated frogs could make jumping motions in response to a signal on their upper spine. The Galvanis eventually published results from their follow-up experiments on how frog legs respond to electricity in 1791. The electrical force that Galvani discovered led to the term "Galvanic force" and eventually to the common English term "galvinize."

In Galvani's experiments, he showed that applying an electrical current to a frog muscle could induce that muscle to contract as long as a nerve was in contact with a metal conductor. He thought that an "animal electricity" must originate in the brain and travel through nerves to the muscles. Although his basic theory was correct—that nerves carry electrical impulses to the muscle—Galvani came under attack from fellow physicist Alessandro Volta (1745–1827) for his theory that animals themselves could be the source of electricity.

On examining Galvani's experiments, Volta didn't see the need to endow the human brain with a source of animal electricity. Instead, Volta argued that the electrical signal came from the differences between two metals, which scientists at the time agreed made more sense than Galvani's animal electricity explanation.

Eventually, Galvani, or perhaps his nephew, published an anonymous set of experiments in which they proved that animal muscles could be triggered to contract by electricity that does not come from dissimilar metals, as Volta predicted. Because of Volta's international acclaim after developing the battery, his argument against an inherent animal electricity was upheld in the scientific community, though Galvani did end up being correct in denying that dissimilar metals lie at the root of a nerve signal. This work by Galvani and Volta had a lasting effect outside biology and is what eventually led to the battery, telegraph, electric light, and the telephone.

The argument between Galvani and Volta kicked off a sudden scientific interest in frog legs. In the next eight years, more than fifty publications emerged describing various results of combining frog legs, electricity, and differing types of metals. Many of these experiments were with the intention of bringing the muscles to life or endowing the muscle with its own animal electricity. It is no surprise that Mary Shelley (1797–1851) wrote *Frankenstein*, in which a scientist brings a dead person to life by applying electricity, soon after this flurry of experiments with electricity.

Among the publications during this time was one by the naturalist Friedrich Heinrich Alexander von Humboldt (1769–1859), who bridged the gap between Galvani and Volta with experiments released in 1797. He acknowledged that the electrical charge between two different metals could trigger a muscle to contract. But that same muscle would also contract when touched by a nerve elsewhere in the body. His view that electricity could come from metals or be generated by animals did end up being correct, but Volta's fame prevented von Humboldt's conciliatory idea from taking hold for more than about forty years.

Regardless of where and how electricity originates in an animal's body, theories abounded about how muscles used that electricity to contract. One theory came from the Italian physician Eusebio Valli (1755–1816). In 1793, Valli published his theory that when muscles were relaxed they had a positive charge on the surface and a negative charge inside. During a contraction the electricity discharged, leaving the muscle without any electrical charge. The muscle then had a latent period during which it could not contract until charges once again built up on the inside and outside of the muscle.

Another theory, this one from Leopoldo Nobili (1784–1835) in 1827, suggested that the electrical current was due to a temperature difference between the nerve and muscle. He predicted that the nerve cooled more quickly than the muscle, establishing a "thermo-electric" gradient. This idea seemed to stump researchers at the time, who didn't comment again on electricity in muscles until 1838 when Carlo Matteucci (1811–1868) published experiments showing, once and for all, that a difference in electrical current does exist between nerves and muscles. Matteucci also coined the term "tetanize," to mean the sustained contraction of a muscle. This word is most commonly used today in reference to the disease tetanus, or lockjaw, in which the muscles of the body contract tightly.

The Nature of the Electrical Charge in Muscles

In terms of muscle research, the late 1700s and early 1800s were taken up almost entirely with a debate over whether animals could produce their own electricity, and if so how and what effect it had on the muscles. Once Matteucci put an end to that discussion with his 1838 publication, most researchers were convinced that animals could produce an electric current that went down the nerves and stimulated the muscles to contract. From here, the enduring question was how that current was generated and how it stimulated the muscle.

During the mid-1800s, all of science took a larger part in the popular imagination. Intellectual salons discussed scientific theories along with literary publication. It was in this setting that Charles Darwin (1809–1882) published *The Origin of Species* in 1859, forever changing the discussion—if not everyone's perception—of the place humans occupied in the universe. The scientific community during this time put a larger emphasis on experimentation, moving away from publications in the past that were based on ideas and general observations.

It was in this environment that Emil du Bois-Reymond (1818–1896) began working on the problem of electrophysiology. Du Bois-Reymond went on to be one of the best-known names in muscle research. He became interested in this area of research after reading works by Matteucci in 1841. At this time, Matteucci had noticed that after contracting for a long period of time, the muscle contained less current. He later went on to contradict this finding, then to contradict his contradiction. Despite his confusion over whether the muscle gains or loses current during a contraction, he did make one important discovery involving the transmission of an electrical impulse. Matteucci found that when the nerve from one limb is connected to a second limb, and that second limb is stimulated to contract, the first limb will contract as well. He attributed this result to "nervous forces" in the second limb.

After reading Matteucci's work, du Bois-Reymond tested the idea that muscles lose their charge as they remain contracted. Publishing in 1848, du Bois-Reymond showed that the negative charge, or "negative variation," took place in both the muscle and the nerve when the muscle was contracted. This negative variation that he described is now known as the action potential that occurs when nerves stimulate muscles to contract. During this action potential, the nerves and muscle do briefly lose their positive and negative charge and nerves cannot fire. For more information on the action potential in nerves, see the Nervous System volume in this series.

The depleted charge that du Bois-Reymond described could only take place if a muscle at rest had a charge. He argued that relaxed muscles always contained a charge and that injured muscles—such as what researchers usually studied in dissected frogs—simply had a greater amount of that resting charge. This question of whether or not resting muscles had an electric current continued to be debated for many years. It was clear that an electrical stimulation was needed for muscles to contract and that a contracting muscle could be the source of an electric current that makes another muscle contract. But the nature of that electric current eluded the early muscle researchers for many years.

One of the key people who argued against du Bois-Reymond's theory was Ludimar Hermann (1838–1903), who had studied under du Bois-Reymond early in his career. Hermann felt that injured muscles had a current but normal resting muscles did not. For contradicting the idea that muscles at rest have a charge, du Bois-Reymond had Hermann thrown out of his lab space at the Berlin Institute. Hermann completed his work in a lab built at his parent's printing press, and he was eventually invited back into a proper laboratory at the University of Zurich. As with the argument between Volta and Galvini, du Bois-Reymond and Hermann were both partially correct. A resting muscle does not have any current along its surface, as Hermann predicted. However, the individual fibers within the muscle do carry positive and negative charges.

A fairly accurate theory about the electrical nature of muscles didn't enter the scientific debate until the late 1800s, when Julius Bernstein (1839–1917) elaborated on one of du Bois-Reymond's theories. He had argued that the nerves and muscles were made up of particles with a positive end and a negative end. These particles lined up side by side, with all their negative ends facing one direction and the positive ends facing the other. These sheets of particles made up the membranes of nerves and muscles, creating a negatively charged interior and positively charged exterior. In fact, this idea was close to being correct. Rather than polarized molecules lined up end to end, a muscle or nerve cell has positive molecules on one side of the cell membrane and negative molecules on the other.

The electrical current in injured muscle that du Bois-Reynolds and others had noticed occurred when the membrane surrounding a muscle fiber breaks and the positive and negative charges mix. In a series of experiments, Bernstein and others noticed that if they set up a glass tube with positive molecules at one end and negative molecules at the other with a somewhat permeable membrane in between, an electrical charge developed while the molecules drifted back and forth across the membrane. Once the two molecules were equally divided on both sides, the current disappeared. Bernstein and others correctly deduced that a similar event was taking place within an injured muscle.

With this set of experiments, the scientific community reached some consensus on the nature of how muscles generate electricity. From here they could move forward with research into the nature of how that electricity triggered a muscle to contract and how the muscle got energy for that contraction.

Energy Dynamics in Muscle

In order to understand how muscles do the work of heavy lifting or regular contraction, it was first necessary to understand what is now known as the first law of thermodynamics. According to this law, energy can change form—it can be stored as sugars, expended as work, or released as heat—but energy cannot be created or destroyed. In a muscle, energy comes from either fats or sugars (see Chapter 2 for more information about how the muscles make energy for contraction). That form of energy is converted into ATP, which is simply another way to store energy. But not all the energy in sugar is converted into ATP—some of the energy is lost as heat, which is why people heat up when exercising. Muscles use energy stored in ATP to lift heavy objects. The energy in the original sugar or fat molecule was converted into movement and is now stored in the heavy object that would in turn release more energy were it to fall.

The first law of thermodynamics was first discovered by a German physiologist named Hermann von Helmholz (1821–1894), who first described his work in 1847. Even in his earliest work, Helmholz was a great believer that nature could be understood. He wrote, "The final goal of the theoretical natural sciences is, therefore, to discover the ultimate and unchangeable causes of natural phenomena." He was convinced that the idea of a vital force that animated people, which many physiologists of his day still believed, should be abolished from science. All of biology could be understood through studying the laws of physics and chemistry. He had not taken a single math course during his higher education, yet by his rigorous science he came up with a fundamental law that underlies the entire study of energy whether it is in physics, chemistry, or biology. Helmholz was only 26 at the time.

Being a physiologist first and foremost, Helmholz proved his law of thermodynamics by studying muscles. He reasoned that heat produced by muscles must come from energy in the body. Because he did not believe in a vital force that could generate energy spontaneously, that heat must come from the energy stored in food. To prove that contracting muscles are what produce the body's heat, Helmholz created a new device that is now known as a thermocouple. It was carefully insulated from outside temperature changes and had room on the inside to contain a frog thigh muscle. Sensors set up inside the apparatus could detect that tiny amount of heat released in a single contraction of that muscle. The physicist James Joule (1818–1889), working at around the same time as Helmholz, worked out how much energy was needed to generate a certain amount of heat. This calculation had not yet been established, leaving it for later researchers to calculate how much energy the muscle used by measuring the heat released. Because of this work, scientists now measure energy in units called Joules. One Joule is the amount of energy it takes to lift a 3.5 ounce (100g) object

1.1 yards (1 meter). Helmholz was the first to notice that after repeated use, an acidic substance builds up in muscles. Many years later, this acid substance—which we now know to be lactic acid—would be incorporated into a coherent idea about how muscles generate the energy to contract.

Rudolph Heidenhain (1834–1897) followed Helmholz's work by looking at how much heat muscles release when doing different levels of work. He attached the muscle in a thermocouple to a weight, then stimulated the muscle to contract and lift the weight. When he doubled the weight, the heat released by the muscle more than doubled, and when he tripled the weight, the heat released far more than tripled. He was the first to discover that muscles become less efficient as they do more work. When lifting light weights, muscles transfer quite a bit of energy from food into work and only a small amount into heat. When lifting heavier weights, the muscle transfers some of the energy from food into work but releases much more as heat. It turns out that muscles—like cars and other machines—are relatively inefficient at converting stored energy (gas in the case of cars) into work and become less efficient the more work they are asked to do.

While Helmholz and Heidenhain were publishing their observations of muscles heating up during contraction, other physiologists were theorizing about how that heat related to muscle contraction. One theory went that the muscles were made of a material that got shorter when heated. The model for this theory was catgut, which grew shorter when heated and longer when cooled. If this model were true, then the muscles would convert energy in food into heat, and that heat would cause the muscle to shorten and do work. It was Adolf Fick (1829–1901) who proved this idea wrong. In 1869 he argued that energy in food was transferred directly into the energy needed to do work. In the end Fick's theory was correct, though he knew nothing about ATP.

Muscle Chemistry

Since the 1600s various researchers had tried to understand the chemistry of how muscles contract. These studies didn't go far, in part because the techniques for studying biological chemistry were in their early stages. Among the early theories of muscle contraction was one stating that salts in the muscle mixed with components of the blood to create a small explosion. This explosion caused the muscle fibers to swell around the circumference and contract. Another early theory was that nerve signals cause chemicals in the muscle to split and activate the muscle to contract.

During the 1700s physiologists began making discoveries that would help guide later researchers in understanding the true nature of muscle contraction. Matteucci, among others, found that the muscles continue to release carbon dioxide even after the oxygen supply had been cut off. This result led Hermann to create a theory involving the "inogen" molecule.

Inogen, according to Hermann, was a precursor to both lactic acid and carbon dioxide, both byproducts of muscle contraction that physiologists had discovered by that time. Muscle contraction broke the inogen molecule into lactic acid and released a carbon dioxide. The lactic acid was later recombined to create a fresh supply of inogen. Eduard Pflueger (1829–1910) built on this theory, adding an oxygen molecule to the inogen. This stored oxygen explained where the molecule came from when the oxygen supply had been cut off.

It wasn't until the late 1800s, when Fick was arguing that muscles convert energy directly into work, that researchers set aside the theoretical inogen molecule. An initial attempt to explain contraction without inogen came from Johannes Gad (1842–1926), who argued that muscles use energy from lactic acid to contract. Oxygen was later added to regenerate the lactic acid for a future contraction—this process of adding oxygen to lactic acid is what he suggested released heat. Adding another piece to this puzzle, Max von Frey (1852–1932) announced in 1880 that muscles use most of their oxygen after the work has already been done, and he found that lactic acid was produced as oxygen was used up.

Finally, in 1907, Walter Fletcher (1873–1933) and Frederick Hopkins (1861–1947) did a set of experiments to show that when glycogen disappears, lactic acid appears in corresponding quantities. Thus, glycogen breaks down to fuel muscle contraction, releasing lactic acid as a byproduct. They also realized that carbon dioxide came from the breakdown of lactic acid rather than from processes that use oxygen. This finding explains why carbon dioxide is released even in the absence of oxygen. Building on this, Archibald Hill (1886–1977) and Otto Meyerhof (1884–1951) shared the 1921 Nobel Prize for their work on the relationship between heat production and muscle contraction, and for discovering the relationship between oxygen use and lactic acid production.

Muscle Response to Stimulation

When a person decides to contract a muscle—to pick up a glass, for example—the muscle can contract either partway to hold the glass or farther to bring the glass all the way to the lips. Clearly, when a muscle gets a signal to contract, it does not have to contract fully. Today, people know this graded response comes from the number of muscle fibers that are called upon to contract. Only a few fibers contract for a small motion, while many contract for a larger motion.

During the late 1800s, Fick, among others, worked on this problem of how muscles produced a graded contraction. One widely accepted theory held that the strength of a nerve stimulation is what determined the strength of the contraction. With this theory, a strong nerve stimulus gives rise to a strong muscle reaction.

Studying heart muscle, researchers began to grasp the all-or-none mechanism for muscle contraction. However, the researchers didn't feel that their recently discovered principle applied to skeletal muscles. Fick was alone in thinking that skeletal and heart muscle may both be restricted to an on/off response to stimulation. He wrote that in skeletal muscle, each nerve impulse gives rise to "either a maximal contraction or no contraction at all." However, Fick was more interested in his other work, and did not pursue this ultimately correct assumption.

It wasn't until 1902 that Francis Gotch (1853–1913) finally proved the all-or-none response in skeletal muscle fibers. He stimulated one branch of the nerve going to a frog's thigh muscle and found that only one part of the muscle consistently contracted. When he stimulated a different branch of the nerve, another portion of the muscle consistently contracted. He correctly theorized that a graded muscle response comes from how many muscle units are stimulated.

A colleague of Gotch, Keith Lucas (1879–1916), refined Gotch's initial idea. Lucas studied a small muscle in the frog that has only seven to nine nerves going to it. He found that as he slowly increased the electrical signal going to the muscle, the strength of the contraction went up in seven to nine distinct increments. Each of these incremental increases in strength came when a new nerve signaled its motor unit to contract. Once the muscle reached its peak contraction, additional stimulation had no effect. Although this evidence seemed conclusive to many researchers, they still had not proven that the individual fibers had an all-or-none response to being stimulated.

In the end, it was Frederick Pratt (1873–1958) who, in 1917, proved once and for all that muscles either contract or do not contract, and cannot modulate the amount of their contraction depending on the strength of the nerve signal. Pratt sprinkled mercury droplets on a single muscle fiber. He then stimulated individual muscle fibers to contract using tiny electrodes that he could insert into the fibers. When he photographed the muscles, Pratt found that the mercury droplet traveled in discrete steps, as if the muscle contracted in a stepwise fashion rather than in a continuous graded fashion. This experiment also showed that the fibers in skeletal muscle—unlike cardiac muscle—are insulated from each other. Stimulating one fiber to contract did not affect neighboring fibers.

The Sliding Filament Theory of Muscle Contraction

Despite impressive progress in understanding how signals travel through muscles and how muscles release heat, physiologists were stymied by the problem of how muscles go about contracting. What was it that caused muscle fibers to grow both shorter in length and wider in circumference?

Through the early 1900s, several competing theories were debated as to how muscles produce lactic acid and carbon dioxide in the process of contraction. One theory held that the long protein strands within a muscle were extended in a relaxed muscle, but then folded back on themselves in a contracting muscle. This folding is what accounted for the muscles getting larger in circumference during contraction. The proteins in this theory were held extended by electrical charges along the protein chain. When a muscle received a signal from a nerve, the proteins strands were thought to release lactic acid, neutralizing the charge and allowing the muscles to fold back on themselves. This rather elaborate theory held until 1930, when experiments showing that muscles could contract without producing lactic acid were published. This finding immediately followed the 1929 discovery of ATP in muscles.

Even with these pieces in place, the full picture of what products are used and produced in a contracting muscle remained unclear. Over the next twenty years, a series of discoveries helped resolve that picture. In 1934, creatine phosphate had already been identified in muscles. Karl Lohmann (1898–1978) worked out the relationship that creatine phosphate is only used up as ATP is produced. This clarified the role of creatine phosphate as a way of storing high-energy phosphates to replenish ATP during short bursts of exercise.

In 1939 researchers discovered a protein (which they called myosin) that could convert ATP into ADP, releasing energy. At this time, electron microscopes had not yet been invented, so the exact structure of skeletal muscle wasn't known. Even without that tool, in 1943 the original myosin was discovered to be two proteins—actin and myosin—that are bound together. Releasing the two proteins, it was discovered, requires ATP.

At this point, scientists had the information they needed to come up with an accurate model for muscle contraction. In the end, groups of researchers independently came up with the correct theory, both with the help of the newly discovered electron microscope. With this new tool, researchers could see the fine structural detail of individual muscle fibers, including the sarcomeres, bands of actin and myosin, and regular Z bands dividing the sarcomeres. Looking at samples from relaxed muscle and contracted muscle, they could see that the sarcomeres shortened during a muscle contraction and the bands of actin and myosin also overlapped to a greater extent.

In 1954, Hugh Huxley (1924–) and Emmeline Hanson (1919–1973) published a paper at the same time as Andrew Huxley (1917–) describing their sliding filament model of muscle contraction. Andrew Huxley (no relation to Hugh Huxley) went on to develop a mathematical model for how the filaments slide past each other. He continued this work, learning how myosin uses energy from ATP to attach to actin, flex, and release, ratcheting its way along the actin filament. By 1971, Andrew Huxley published his analysis of the amount of force the myosin heads generate with each cycle of attaching and releasing from actin. Adding the individual forces of all the myosin in a muscle equals that total force generated by that muscle.

Exercise Physiology

Throughout most of history, the study of muscles was limited to trying to understand how the muscles function in isolation. But for many people, the only time they notice the muscles is during exercise—or the day after exercise, in some cases. For thousands of years civilizations had prized fast runners or strong men, and had recognized the phenomena involved in exercise such as the burning sensation, fatigue, cramping, and soreness following exercise. With that in mind, it is somewhat surprising that the first publication on the role of muscles in exercise wasn't published until 1880. At this time, the understanding of muscles in exercise was so limited that even the author admitted in the text that some of the theories may not be accurate.

One of the earliest researchers to study exercise was Nobel Prize winner Archibald Hill (1886–1977). Although he worked in the 1920s, before the discovery of the sliding filament model of muscle contraction, Hill was able to calculate relationships between oxygen consumption, lactic acid production, heat, and workload. He was an avid runner who coined the term "oxygen debt" to describe his own fatigue after track meets. With the help of John Haldane (1860–1936), who had developed an apparatus for measuring oxygen use during exercise, Hill pioneered efforts to study athletes while exercising. This type of equipment, now attached to sophisticated computers, is still used to study exercise physiology.

Some of the most important contributions to exercise physiology came from the Harvard Fatigue Laboratory (HFL), founded in 1927. Members of this lab realized that muscles respond differently when at high altitudes or in extreme temperatures, and they carried out their experiments in remote locations as well as in the lab. Members of HFL were also the first to realize that muscle physiology changes with age. Studies at this lab by Sid Robinson first looked at oxygen use and heart rate in people ages 6 to 91. The HFL closed in 1947, when universities around the world had begun to realize the importance of exercise physiology and welcomed the research within established departments.

For much of the past fifty years, the field of exercise physiology has focused on measuring attributes of the whole body during exercise. With muscle samples taken before, during, and after exercise, combined with data about workload, oxygen consumption, carbon dioxide production, and blood sugar and fat levels in the blood, physiologists have been able to piece together a detailed understanding of how the body reacts to exercise different types of exercise.

HEART MUSCLE

During the Middle Ages, scholars relied on the Greek writings for their understanding of physiology. In particular, Aristotle and Galen guided theories of how the body functioned. Around the thirteenth century, the idea formed that the heart was the *membrum principale*, the principle organ. It took the central position in the body, with all the other organs grouped around it like planets around the sun. In this position, the heart played the central role in animating the body.

In the mid-1600s, research into how the heart functioned began to draw a more accurate picture of the heart's role in the body. Swammerdam, who found that skeletal muscle volume doesn't expand during contraction, did the same type of experiment with the heart and reached same conclusion in heart muscle as in skeletal muscle. Swammerdam also noticed that the heart could contract, though somewhat irregularly, even without an attached nerve. This ability to contract without a nerve distinguished the heart muscle from the skeletal muscle, which lost all ability to contract when the nerve was severed. George Baglivi (1668–1708), who worked soon after Swammerdam, extended this work and noted that even isolated pieces of heart muscle could "tremble and oscillate" on their own.

Working in the same time period, Borelli—who theorized that muscles expanded and contracted much like a lattice gate—applied his theory of muscle contraction to the heart. He theorized that when the muscles contracted, the fibers swelled and pushed into the heart chambers, making the chambers smaller and expelling blood.

Although he did not contribute to understanding how the heart muscle contracts, Marcello Malphighi (1628–1694), who was an expert at preparing tissue samples to preserve their structure, noticed the disorganized, almost coiled direction of the individual muscle fibers. It is the branching nature of the heart muscle fibers that gives the heart a wringing contraction to squeeze out blood. Malphighi published his work in 1657. About twenty years later, Anton Leeuwenhoek turned his expertise with a one-lens microscope to cardiac muscle and confirmed the net or weblike interconnections of the fibers.

Boerhaave, the Enlightenment-era physician with a mechanistic view of the muscles, also put his mechanical mind to work on the heart muscle. He concluded that fluid from the nerves inflated the heart muscle, causing it to expand and push blood into the arteries. The dilated arteries triggered the nerves to stop emitting the fluid, causing the heart to relax and allow new blood to enter. Body heat, he thought, resulted from friction between blood particles and the walls of the heart and blood vessels.

Adding to this growing understanding of how heart muscles contract, experiments published in the 1760s showed that by pricking one single nerve fiber, the entire heart could be triggered to contract. The researchers were beginning to understand that the heart muscle fibers spread the signal to contract, while skeletal muscle fibers contract singly.

Von Hallers had an unusual view about how the heart contracts. He believed that only skeletal muscle needed a nerve signal in order to contract. In contrast, the heart was the most irritable organ and could beat in the absence of any nerves. He felt that the heart muscle was triggered to contract by the incoming blood. Once that blood was expelled, the heart relaxed. An experiment by Whytt, however, demolished this theory. Whytt tied off blood vessels that bring blood to the heart and found that the heart continued beating even in the absence of either incoming blood or a nerve connection.

The next step in understanding how the heart muscle contracts came during the time of active research in how muscles respond to electricity. Remembering Valli's argument that the skeletal muscle goes through a latent phase while waiting to recharge, Felice Fontana (1720–1805) felt this theory could also explain the heart's contraction. When the heart relaxes it goes through a phase—now known as the refractory period—where it cannot contract when stimulated. Fontana felt that this phase must correspond with that period of time when the muscle was regaining its charge. Fontana's publications were the first to hint at the "all-or-none law," in which the heart can either contract or not contract, but cannot contract partway. Despite having worked out the all-or-none law of muscle contraction in the heart, it was many years before those same researchers realized that the principle held true in skeletal muscles as well.

In 1871, researchers were learning that in cardiac muscle, the amount of stimulation the muscle received determined the strength of the muscle contraction. However, a researcher named Henry Bowditch (1840–1911) noticed that this trend did not hold true with cardiac muscle. He found that when the heart muscle received a stimulation it either contracted or did not contract, and that each contraction was equally strong. At this time, the researchers had learned that heart muscle fibers could spread the nervous signal. They assumed correctly that if one muscle fiber was stimulated to contract, all fibers in the heart would contract.

Early in the twentieth century, three researchers each led efforts to study how the heart spreads the signal to contract. Studying these internal signals, as well as the pressure within the heart during a contraction, became much easier after the invention of the oscillograph, a version of which is still used to diagnose heart disease. Werner Forssmann (1904–1979), André Cournand (1895–1988), and Dickinson Richards (1895–1973) won the Nobel Prize for this work in 1956.

By this time, researchers had a good idea of how the heart muscle contracts and spreads signals. Most modern research on the heart muscle has to do with treating diseases that occur when the heart muscle doesn't function properly.

SMOOTH MUSCLE

From as early as Galen, people recognized smooth muscle as a contractile tissue, but they were generally not considered to be actual muscles. It wasn't until the beginning of the 1700s when Baglivi distinguished the smooth muscles from skeletal muscles under a microscope that smooth muscles were considered to be a unique muscle type. He could see that smooth muscles contained a sheet of cells, all of which could contract, whereas skeletal muscle contained long, striped fibers. He accurately concluded that the more membranous smooth muscles were used for maintained contractions rather than quick movements.

Throughout most of the time that skeletal and heart muscles were a focus of research, smooth muscles were largely overlooked. What was learned about muscle contraction was generally attributed to smooth muscles as well, without much additional research. Smooth muscles began to get more attention when it became clear that they played a role in clogged arteries or digestive diseases. Today smooth muscles are studied primarily in conjunction with those diseases rather than on their own as a unique tissue type.

6

Current Approaches to Understanding the Muscular System

Throughout most of scientific history, researchers occupied themselves with trying to understand the basic mechanisms of how muscles contract. Today, that work has contributed to a good overall understanding of the sliding filament model of muscle contraction, a detailed view of how muscles generate ATP under different conditions, and an understanding of how muscles respond to stress and exercise. Despite this seemingly complete picture of muscle research, many laboratories around the world have active research programs in muscle biology. For the most part, these researchers are trying to understand the exact molecular nature of how muscles function under different conditions.

Biologists who study muscles use many of the same sophisticated molecular tools as other researchers. They use stains to highlight proteins within muscles, study muscle cells in the lab dish, and study mice that lack particular genes involved in muscle biology. In addition to these standard tools, exercise physiologists rely on muscle samples taken before, during, or after exercise to learn about the effect of various exercise conditions on the muscles.

CONTRACTION

Dynamics of a Contraction

The basic model for how muscles contract is widely accepted among physiologists. However, there are still some unanswered questions about exactly how much power each actin-myosin connection can generate. In other fields of research, interactions between proteins such as actin and myosin can be studied in isolation in a test tube or lab dish. This type of experiment gives a clear picture of exactly how those proteins interact without the distraction of other interactions within the cell. This type of experiment hasn't been possible when studying proteins within the muscle. The various proteins within a sarcomere exist in a highly regular lattice of actin and myosin with many interconnecting proteins holding these filaments in place. Any studies that don't investigate muscle contraction within this framework will give an incomplete picture of how the muscle normally contracts.

To get around this problem, some researchers address questions about muscle contraction using single, isolated muscle fibers. In some ways, these experiments mimic those experiments being done in the early twentieth century using whole frog muscles. The difference is that researchers now study isolated muscle fibers, which don't have the associated nerves, blood vessels, and connective tissue that whole muscles have. Modern experiments also use much more sophisticated machinery to precisely regulate the work being done by the muscle and to carefully monitor any changes in the muscle length.

Some experiments using a single muscle fiber have focused on the dynamics of how a muscle contracts. By quickly increasing or decreasing the weight being lifted by the fiber, physiologists have noticed that the muscle seems to have some elasticity. When the weight is reduced and the fiber is able to lengthen and then the weight is suddenly added back, the muscle contracts with greater strength than before. This is thought to be because the myosin heads stretch somewhat, then recoil, much like snapping an elastic band.

Some researchers have looked at sarcomere length while the isolated fiber is made to lift more or less weight. To do this, they shine a laser light through the muscle fiber. Actin and myosin bands within the muscle scatter that laser light into distinct patterns. Those patterns can then tell how extended or contracted a given sarcomere is within the muscle. This type of experiment has revealed that when weight is added to one end of a suspended muscle, the signal to contract travels to the other end of the muscle at 180 meters per second. Researchers also found that the signal to contract with more force took place before the sarcomeres had actually shortened.

One complaint about using single fibers to study how muscles contract is that there is no way to control conditions within that fiber. Being able to modify what chemicals are in the muscle fiber and seeing how the fiber contracts differently under those conditions would help clarify how muscles normally contract. One way to overcome this problem while still keeping the muscle proteins in their intact form is to use muscle fibers that have had their surrounding membrane removed. These are sometimes called **skinned fibers** and consist of only the protein network of actin and myosin without the surrounding membrane.

The same type of experiments can be done with skinned fibers as with single, intact muscle fibers. The difference is that with skinned fibers, researchers can expose the fibers to different types of compounds to see if there are any changes in the dynamics of how the muscle contracts or how the sarcomeres change length. This type of experiment is what helped researchers understand what role calcium plays in muscle contraction. As mentioned in Chapter 1, when a muscle receives a signal to contract, calcium floods into the muscle fibers from the sarcoplasmic reticulum. Calcium binds to the troponin, causing the troponin-tropomyosin complex to change shape and reveal a site where myosin can bind to actin.

To understand the role calcium plays in a contracting muscle, researchers put the fibers in solutions with high or low concentration of calcium, then looked at how those solutions altered the contraction strength and sarcomere shape. This type of experiment is still being used to understand all the molecular changes that take place within the muscle fiber during a contraction.

ATP Use during a Contraction

Despite its critical role in muscle contraction, researchers still have many unanswered questions about how myosin breaks ATP into ADP, how quickly that reaction takes place, and at exactly what point during the contraction cycle myosin splits the ATP. To get at these questions, researchers have several artificial versions of ATP that they can use to study how ATP is converted into ADP in a muscle fiber.

CAGED ATP

The term "caged ATP" refers to a molecule of ATP that is attached to an additional molecule that prevents the ATP from being used normally by the cell. When that caged ATP is exposed to a signal such as a certain type of light, the caging molecule is released and the ATP can be used normally.

Caged ATP is often used in conjunction with skinned fibers in different concentrations of calcium or other chemicals. These types of experiments give an even more precise picture of the events that take place during a muscle contraction. For example, in one set of experiments researchers removed all calcium from a skinned fiber and added caged ATP. When they flashed a light to remove the cage, the fiber initially stopped contracting. This occurred because the myosin bound ATP and released from the actin binding site. Without any calcium in the fiber, the myosin heads could not reattach to the actin and continue the contraction. The rate at which the muscle relaxed gave researchers a better idea of how quickly myosin heads bind ATP.

FLUORESCENT ATP

As with caged ATP, fluorescent ATP contains an additional molecule attached to the ATP. Rather than preventing the ATP from working, the fluorescent molecule allows researchers to track the ATP within the cell. These fluorescent molecules come in a range of colors that are visible under a microscope. They fluoresce only when the molecule is exposed to a particular wavelength of light, which the molecule absorbs as energy, then releases as fluorescent light.

Fluorescent ATP is often used to monitor how long it takes for ATP to be used by the cell. For example, if a stripped fiber is in a solution with fluorescent ATP, it shows up as fluorescent stripes under the microscope where the ATP is attached to myosin heads in the sarcomere. By having caged ATP also present in the solution, flashing a light will trigger the caged ATP to release and take over as the fluorescent molecule is used by the myosin head. Under a microscope, the fluorescent bands disappear as the fluorescent molecule is used up and replaced by the caged ATP. The rate at which the bands go away indicates how quickly the myosin used the fluorescent ATP.

New fluorescent molecules are becoming available that make the fluorescent ATP behave as much as possible like regular ATP. These molecules, as well as fluorescent molecules that fluoresce in colors that are less likely to disrupt surrounding tissues, have opened up some new areas of research. One of these is understanding new forms of myosin that have been discovered. There are two general types of myosin, fast and slow. These break ATP faster or slower and are part of what makes a muscle fiber either fast twitch or slow twitch. In fact, there are several different types of fast or slow myosins that all have slightly different properties. With the improved fluorescent ATPs, researchers can do more precise experiments looking at how quickly myosin breaks ATP into ADP. From these experiments, they hope to learn more about the twelve different forms of myosin that have been identified so far.

Other experiments involving fluorescent ATP have to do with exactly how much ATP is necessary for a muscle fiber to contract a certain distance. This should be an easy experiment—put a known number of ATP molecules in a solution with a stripped fiber and see how far that fiber contracts. Unfortunately, it is very hard to know precisely how much ATP is in a solution. Given this problem, researchers have been debating the most accurate way to do the experiment. With newer types of fluorescent ATP that allow researchers to actually see an ATP molecule being used, along with more accurate ways of measuring fibers, there may soon be a precise answer to the question of how much ATP is required for a given contraction.

Calcium Use during a Contraction

Calcium stored in the sarcoplasmic reticulum floods into a muscle fiber during a contraction. Exactly how the signal reaches the sarcoplasmic reticulum to release calcium, how much calcium is released, and how quickly it is taken up again by the sarcoplasmic reticulum are all areas of ongoing research. Two groups of chemicals have helped researchers keep track of calcium within a cell. These are calcium indicators, which turn colors depending on the concentration of calcium, and caged calcium molecules that function in a very similar manner to caged ATP.

CALCIUM INDICATORS

The most widely used calcium indicators fluoresce in the presence of calcium at different concentrations. A range of indicators are available in different colors and with sensitivities to different concentrations of calcium. Researchers can soak a muscle fiber in these indicators, then stimulate the fiber to contract. Under a microscope, the calcium indicator fluoresces brightly within a few milliseconds after the signal to contract, indicating that calcium has flooded into the cell. The fluorescence then slowly disappears as calcium is transported back into the sarcoplasmic reticulum.

CAGED CALCIUM

As with caged ATP, caged calcium consists of calcium bound to a "caging" molecule. A flash of light separates the caging molecule, releasing calcium in an active form into the cell. Caged calcium can be used to carefully control when a muscle fiber receives a burst of calcium. These molecules can be used in combination with calcium indicators to see where calcium is used most quickly within a muscle cell and how quickly it is taken back up by the sarcoplasmic reticulum. These can also be used to control how much calcium is present in a cell. For example, researchers could have a low concentration of a caged calcium within a muscle cell and investigate how strong the contraction is under those conditions.

Interaction of Actin and Myosin

STUDYING PROTEINS IN SINGLE CELLS

Each organism, including humans, has many different genes for both actin and myosin. These genes make different forms of the protein that is used by slow- or fast-twitch fibers; by heart, smooth, or skeletal muscles; or by muscles in different phases of development. In most cases, these proteins are very similar, with just a few slight modifications.

To understand how researchers study these different forms of actin and myosin, it is first necessary to understand how proteins differ from each other. All the proteins of the body are made up of twenty amino acids arranged in a particular order, like beads on a necklace, to make proteins with unique properties. By swapping a few amino acids, the protein can end up with very different properties. A small switch such as replacing one large amino acid for another may have relatively little impact. But replacing an amino acid with one that has dramatically different properties can disrupt the function on the entire protein. In the case of different forms of actin and myosin, usually about three amino acids may be different between the forms. These make the proteins use ATP at a different speed or bind with a different strength to other proteins in the sarcomere.

When two forms of a protein differ by three amino acids, it can be hard to tell how each small change adds up to the overall differences between the two protein forms. To better understand each individual change, researchers have come up with a way to study each incremental change. They create a gene sequence for one version of the protein, then make one of the modifications that exists in another version of the protein. They then insert this gene sequence into a muscle cell in a lab dish. That cell then makes the slightly modified form of the protein.

Researchers can then use fluorescent molecules, caged molecules, and other standard experiments to measure how well that modified protein breaks ATP or binds to other proteins. They can also measure the strength of a contraction with the modified protein. By doing this type of experiment, it is possible to work out how all the small changes in amino acids make the different actin and myosin forms unique.

Another use for this type of experiment is in learning which portions of the proteins are most critical for contraction. Rather than changing an amino acid that normally varies between forms of the protein, researchers modify amino acids that tend to be the same in all forms. They then test the ability of this new protein to function. If, after changing only one amino acid, the resulting actin-myosin complex can no longer cause a contraction or break ATP, they know that amino acid is particularly important in the normal role of the protein. This type of experiment has helped piece together how the various proteins in the sarcomere interact in their complex three-dimensional arrangement.

STUDYING PROTEINS IN ANIMALS

Studying altered proteins in single cells is an excellent way to learn about the dynamics of individual changes on the properties of the protein. However, this type of experiment only hints at how such an alteration would function in an actual muscle. To learn about altered proteins in functioning muscles, researchers use research animals such as mice and fruit flies (*Drosophila melanogaster*).

Both mice and fruit flies have been studied extensively in the lab, and there is a wealth of information about the genetics and biochemistry of these animals available to researchers who study them. Because researchers know exactly how these animals behave and how their muscles normally respond, they can easily measure what happens when the muscles contain an altered form of the protein.

There are many ways to create mice and fruit flies with altered proteins. Whichever way is used, the result is an animal that does not make the normal form of either actin or myosin, and instead makes a form with a specific alteration. Researchers can then study how the muscle functions, how quickly it gets fatigued, and (in the case of the mouse) whether the altered fibers behave as slow- or fast-twitch muscles.

The two approaches—studying cells in a lab dish and studying whole animals—add up to a complete picture of how actin and myosin interact, and which portions of the protein are most critical. Experiments with single cells give precise information about how the alterations change the protein, whereas experiments with whole animals show how the altered proteins change how the muscle works.

Watching Proteins Interact

Within a sarcomere, actin and myosin filaments are tightly packed and hard to observe individually. Instead, most experiments focus on the bulk effect of all those interactions on a muscle contraction. However, some experiments give researchers a way to look at individual myosin proteins moving along an actin filament. This type of experiment takes advantage of fluorescent actin or fluorescent myosin and altered proteins produced by cells in a lab dish.

The fluorescent actin and myosin behave very much like fluorescent ATP. The proteins are attached to a fluorescent molecule that can be seen under a microscope. In one type of experiment, myosin is purified and attached to a glass slide. Actin filaments with bound fluorescent molecules are then washed over the slide surface. Watching through a microscope, a researcher can see the actin filaments being pulled across the slide surface by the myosin. Researchers usually videotape the experiment, then use the video image to measure how fast the actin moves across the slide.

Researchers can compare different normal forms of actin and myosin and compare those with altered proteins. In one recent set of experiments, rather than making alterations to normal myosin, the researchers only made a small portion of the myosin molecule. They then worked out which small portion of myosin was sufficient to move actin across the slide. The remainder of the myosin protein is still important to hold the myosin in place within the sarcomere, but with this experiment researchers now know which region of the protein is directly involved in binding actin and pulling against it during a muscle contraction. To get an even better understanding of this interaction, proteins with alterations in just this region can be used in a similar type of experiment.

An alternative form of this same experiment involves attaching myosin to fluorescent beads and monitoring the beads as they move across parallel actin filaments. The advantage to this experiment is that the actin filaments are aligned, so the beads all move in parallel tracks in one direction, making them much easier to monitor. Myosin on a glass slide is not aligned, causing the actin filaments to move across the slide in a somewhat haphazard fashion.

Using myosin attached to fluorescent beads has been the most efficient way to measure contraction speed of different forms of myosin. As expected, those forms of myosin that break ATP into ADP most quickly in test tube experiments also moved fastest across the actin.

Although both of these experiments are useful for learning how actin and myosin interact, neither gives a good indication of the strength of a contraction or the amount of ATP needed for a given contraction strength. To address these questions, one group of researchers put myosin on a glass needle and watched that glass needle move across aligned actin filaments. By putting a small amount of myosin on the needle and including a very small amount of ATP in the solution, the researchers could see tiny incremental steps as the needle moved across the actin. These steps are thought to be the movement caused by a single flexing myosin head. The number they came up with, 10 nanometers (roughly 1/100,000,000 of a meter), is quite close to what researchers have seen in other experiments.

To address the question of precisely how far myosin pulls actin in a contracting muscle, and with how much force, researchers are configuring variations on the two models of actin sliding across myosin, or myosin moving across actin. They have gotten closer to an answer by attaching the myosin to magnetic beads, then putting the entire experiment in a magnetic field. This provides an easy way to measure how quickly myosin moves the beads across the actin under different magnetic strengths. Combining this with newer fluorescent molecules that allow researchers to see a single actin or myosin protein rather than a conglomeration of proteins could lead to a clear relationship between force and energy.

Viewing the Muscle

Around 1950, a tremendous breakthrough in scientific instrumentation led to the discovery of how actin and myosin slide across each other during a contraction. That breakthrough was the electron microscope. Normal light microscopes can magnify images to the point where some of the structures within a cell become visible, and they were powerful enough for Leeuwenhoek to pick out striations in skeletal muscles. However, they are not powerful enough to view detailed structures within the sarcomere.

Electron microscopes can magnify images to show intricate details within cells and within the sarcomere. Images from electron microscopes show details of the junction between nerves and muscles, mitochondria in large numbers in fast-twitch muscles, blood vessels winding between muscle fibers, and sarcoplasmic reticulum and T tubules within the fibers themselves.

Samples for electron microscopy have to be extremely thin. Because it isn't possible to slice most samples finely enough without damaging the structures, they are usually frozen and embedded in a type of plastic to preserve the structure, then that plastic is sliced for viewing. Because of this step and the conditions within an electron microscope, electron microscopes cannot be used to watch events as they occur. Instead, the slices paint a picture of what was going on within a cell immediately before it was frozen.

In addition to revealing the fine inner structure of the sarcomere, electron microscope experiments can show changes in the structure of a muscle as a result of different conditions. For example, isolated muscle fibers or skinned fibers can be frozen when relaxed or when contracting under different conditions. They could also be frozen immediately after the release of caged ATP molecules or after contracting in different concentrations of calcium. By comparing the results of the contraction strength with how the muscle looked under an electron microscope, researchers can piece together exactly how the muscle contracts under these different conditions.

Electron microscopy has also been useful in combination with experiments using myosin attached to a surface with actin filaments moving over top. In one experiment, researchers were able to see that bridges between the actin and myosin were at different angles when they were frozen after being exposed to ATP versus when they were frozen in the absence of ATP. This same experiment, done with just the isolated subsection of myosin that is responsible for binding to actin, allowed researchers to study that important part of the myosin protein in isolation.

MUSCLE METABOLISM

Measuring Muscle Enzymes

At any given time, a muscle fiber is producing ATP through a variety of means. One way of measuring exactly what is happening within a muscle at a given time or during a given type of exercise involves freezing a muscle and extracting its contents. This can be done on isolated muscle fibers, on isolated muscles, or with samples taken from the muscles during different stages of exercise. The resulting mixture gives a clear picture of how much of a particular molecule is present in the fiber. For example, it can indicate how much creatine phosphate has been used to regenerate ATP, how much lactate has built up in the cell, or how much of different types of enzymes are present in fast-twitch versus slow-twitch fibers.

Freezing a muscle and extracting the contents is an excellent way to get an overall picture of what processes were taking place within that muscle. This type of experiment is what helped researchers understand when creatine phosphate gets used during exercise and how lactate builds up in fast-twitch versus slow-twitch muscles. It is also possible to compare blood lactate levels to lactate in a tissue sample at that same time point to understand when lactate begins diffusing out of the muscle and building up in the blood.

Measuring Oxygen Consumption

The amount of oxygen a muscle fiber uses is directly proportional to how much work that fiber is doing and how the fiber is generating ATP. If the ATP comes from the citric acid cycle and the electron transport chain, then the muscle must use oxygen to produce ATP. One way of measuring how much oxygen a particular fiber uses is through an oxygen electrode, which contains a solution that conducts current in proportion with the amount of oxygen that is present in the sample. The electrode and a whole muscle or isolated muscle fiber are both put in a chamber. Researchers can then manipulate how often or how forcefully the muscle contracts and at the same time record how much oxygen the muscles uses. This type of experiment has helped researchers understand how much oxygen the fiber uses under different conditions. It can also be used to study isolated fast-twitch or slow-twitch muscle fibers to see how those fibers differ.

This type of experiment is much more precise than monitoring how much oxygen a person uses when running on a treadmill or riding a stationary bike. Although both of these experiments are useful for learning how the whole body uses oxygen, they don't allow researchers to investigate how oxygen is used by individual fibers within the body. If researchers quickly freeze these isolated muscles, they can also compare the oxygen use in that muscle fiber with factors such as lactate or creatine phosphate levels to get a better overall picture of how that cell was producing energy.

Measuring Lactate Production

When muscles produce energy anaerobically, they build up lactate within the fiber. This lactate seeps out into the bloodstream when it is present in low quantities, but the fiber actively moves it out once it builds up in concentration. To measure how much lactate a fiber is producing, researchers put the isolated muscle or fiber into a closed container and measured how much lactate is released into the solution.

DEVELOPMENT

Identifying Muscle Precursor Cells

During embryonic development, skeletal, heart, and smooth muscle precursor cells originate throughout the embryo. Researchers have learned where the cells originate by looking for specific molecules made by those cells. The most common molecules to look for are specific types of myosin made only by that cell type. Embryonic myosin, for example, indicates the primary muscle cells moving to the limb buds, whereas the secondary myoblasts use fetal myosin. Likewise, early heart muscle cells use a specific form of heart myosin that can be used to distinguish them from surrounding tissue that looks quite similar.

Finding these unique proteins within an embryo usually involves **antibody** staining. Antibodies are molecules made by the immune system that recognize and bind to very specific molecules. An antibody for fetal myosin, for example, will bind only to fetal myosin and will not bind to embryonic or cardiac myosin. For these antibodies to be useful, researchers must be able to see where they are located. To achieve this, they attach a visible molecule to the invisible antibody.

Most of what researchers know about human development comes from research using mouse embryos. The researchers take a mouse embryo and immerse it in a solution of the antibody. The antibody finds the protein of interest within the embryo, binds, and is visible under a microscope. This type of experiment can highlight early heart muscle cells beginning to form the heart tubes, or skeletal muscle cells migrating out to their eventual locations within the limbs.

One reason researchers know so much more about skeletal and heart muscle development than they do about smooth muscle development comes from the fact that smooth muscle precursors are harder to identify. They use several myosins in common with other muscle types, so it can be hard to distinguish those cells from surrounding skeletal muscle cells. In fact, researchers still aren't sure whether heart and smooth muscle precursors can switch cell types, because they can't be sure that they are seeing exclusively smooth muscle precursors.

This same type of experiment can be used to see what cells are beginning to make muscle-specific proteins such as MyoD. By staining embryos with antibodies to a variety of muscle-specific proteins, they can begin to piece together when those proteins are used during the course of development.

One problem with this technique has been that researchers need to know about a protein in order to develop antibodies to it. Staining with antibodies is a good way to know what role known proteins play, but it cannot identify any new—and possibly very important—proteins.

With the sequencing of the genome, researchers can search for myosin proteins or proteins that are similar in structure to other known developmentally important proteins such as MyoD. By finding these and making antibodies to them, researchers can learn about proteins that had not previously been known to be important in development.

Studying Developmental Regulators

During embryonic and fetal development, a cascade of signals controls how each individual cell matures. One way to learn what these signals do is to block them from occurring or apply them at the wrong time. It isn't possible to carry out this type of experiment in embryos such as mice that mimic human development. Mammals have to develop inside the uterus of the mother, so it is difficult to manipulate their development. However, researchers can remove early muscle precursor cells and manipulate them in a lab dish. This type of experiment doesn't exactly mimic what happens during development, but it can help researchers understand how early muscle cells respond to different signals.

In some cases, researchers may want to know what effect a particular chemical has on a muscle cell. In this case, they can grow the cell in different concentrations of that chemical. Many times in development, the exact concentration of the signal is important for what type of effect it has. For example, a high concentration of a signal may indicate that the cell is near to another structure and should develop one way, whereas cells that sense a lower concentration are farther from the structure and should develop a different way.

In addition to growing cells in different concentrations of a particular signal, researchers can study the effects of a given tissue. This type of work is what helped researchers understand the role the notochord plays in guiding the developing somite. In 1956, Avery and his colleagues placed a small amount of notochord tissue in a dish with muscle precursors. Those then began developing as muscle cells. Some early nerve-type cells play a similar role in directing the development of the heart muscle.

In addition to studying cells in the lab dish, researchers can very carefully control some of the factors in a developing embryo. Many of these experiments are done in chicks rather than mice, because chicks are easier to manipulate during development. These experiments don't absolutely reflect what goes on during human development, but many processes are very similar in birds and mammals.

Researchers may identify a compound in the lab dish that seems to control how muscle cells mature. They can then release some of this chemical into a region of the developing chick to see if it has the expected effect. Alternatively, they may try to block the tissue that is releasing that signal to see if the absence of the signal changes how the muscle cells develop.

Identifying Genes Used by Developing Muscle Cells

Part of understanding how a muscle cell matures is understanding what genes that cell uses during different phases of development. Although all cells have the same DNA, they all make use of different genes. Some genes are only used during crucial steps of development in a few types of cells, whereas other genes are used widely by all cell types. When a gene is being used, it makes a molecule called an RNA. This RNA is essentially a sentence that the cell reads to understand what protein that gene makes. The more active a gene is, the more RNA it makes. With that in mind, researchers look for levels of RNA to indicate which genes are used by a cell.

Until recently, there was only one way to look at RNA levels within the cell. This method could only be used to look for levels of a single RNA and compare that with levels of the same RNA in a different cell or tissue. This was a good way to study cells that had been exposed to different conditions in a lab dish. Researchers could then look at a known RNA to see if the conditions caused the populations of cells to start using different genes and therefore start down a different developmental pathway.

Now researchers have a new, very powerful tool at their disposal for looking at RNA levels. Researchers can buy glass slides that are dotted with thousands of genes—some slides contain all of the known human genes, or all of the mouse genes. These slides, called **gene chips**, have come about thanks to data from genome sequencing projects that have revealed all of the genes in an organism.

To use a gene chip, researchers isolate all the RNA in a cell population. They then attach a fluorescent molecule to that pool of RNA, and they wash the RNA over the glass slide. The RNA then sticks to its gene on the chip. A particularly active gene that made a lot of RNA will make a bright spot on the glass slide. Researchers can then compare the pattern of spots made by two different cell populations or two different tissues to get an overall picture of which genes were being used in each population.

Whereas older experiments could only tell researchers relative levels of known genes, this experiment can highlight genes that researchers did not know were involved in development. By exposing cell populations to different conditions, then carrying out a gene chip experiment, researchers can see how those conditions affected levels of every gene in the genome.

Once they have found a gene that is used at different levels in two populations of cells, researchers can then make an antibody to the protein and look to see where it is made in a developing organism. They can also use the same antibody to monitor the cell populations to see if other conditions also cause the protein to be made. Gene chips are quickly becoming an important way for researchers to learn about how genes are used in different cell types.

Studying Proteins in Living Cells or Animals

Knowing when or where a gene is used doesn't give a researcher much information about the role its protein product plays in normal development.

To understand the normal function of a protein, researchers often prefer to study the protein in isolated cells or in whole animals.

CELL MODELS

Researchers can insert a new gene into cells in a lab dish and have that gene make a new version of a protein. This protein may have a few amino acids different from the normal protein, or it might be in a form so that it is always made rather than only made in response to the appropriate signal. With these altered proteins being made, researchers can expose the cells to known signals and see if their response has changed. For example, if a signal normally causes the cell to begin making a muscle-specific protein, they can monitor to see if their inserted gene changes that normal response. By manipulating individual genes, researchers can learn more about how that gene normally functions.

In recent years, new forms of genes have become available that make a protein with an attached fluorescent section. These fluorescent proteins behave the same as the normal proteins but can be seen under a microscope. This type of experiment is a useful way for researchers to learn more about a newly identified protein. They can also make proteins with alterations and a fluorescent section. These are useful for studying whether alterations in the protein change where that protein goes in the cell.

ANIMAL MODELS

Although researchers can learn a lot by studying proteins in isolated cells, those cells still behave differently than a living organism. For many years now, researchers have been able to create mice that either lack particular genes or that contain altered versions of normal genes.

Mice that lack a functioning copy of a gene are called **knock-out mice**, because the normal gene has been knocked out. In some cases, knock-out mice don't survive because the protein being studied is absolutely essential to the animal's survival. Genes that are involved in development are likely to be deadly when knocked out because the animal can't develop normally. Even in those cases, researchers can study the embryos to see how development goes awry.

One example of this type of experiment is with the protein pax-3, which prevents muscle precursor cells from maturing too early. Researchers had long known about a mouse strain with a mutation that caused offspring to be born with no muscles in the limbs. In these mice, the muscle precursor cells had all matured while still in the somite and never migrated to the limb. Later, after doing experiments that identified the protein they called pax-3, researchers learned that the mouse strain had a mutation in the pax-3 gene. Experiments in the lab dish about the role pax-3 plays were backed up by what they found in the animal model.

In addition to knocking out genes, researchers can insert slightly modified versions of proteins into the mice. In the case of pax-3, researchers could insert a modified version of the pax-3 gene into the mutant mice. They could then look to see if the altered protein was sufficient to make the muscles develop normally. This type of experiment can help researchers understand what portions of the protein carry out important roles in normal development.

EXERCISE PHYSIOLOGY

Exercise physiologists use many standard molecular biology tools to learn more about how muscles respond to exercise. However, they use these tools to answer a unique set of questions.

Muscle Biopsy

The only way to tell whether an exercise regime has changed the muscle physiology is to look at the muscle before and after the exercise, or look at two muscles, one of which went through the exercise and one of which did not. This is easy to do in animals, where the muscle can be removed after the exercise regimen ends. In humans, however, it is harder to examine the muscle for changes. To get a look at the muscle without damaging it, researchers take a muscle biopsy—a small piece of the muscle that is taken out using a narrow biopsy needle.

The biopsy tissue can be studied under a microscope to look at physical differences between trained and untrained muscles, or between muscles that have had endurance versus sprint exercise. A biopsy from the same muscle taken before and after a training regime can also indicate how that type of exercise changed the muscle.

A muscle biopsy is large enough to show the general physiology of a small amount of muscle. It can reveal changes such as the size of an individual fiber, the number and size of mitochondria, the amount of fat droplets used for energy during endurance exercise, or the ratio of fast- versus slow-twitch muscles. It was through this type of exercise that researchers learned basic information about the effects of different training regimes—information that has modified how top-rated athletes train for events.

Muscle biopsies are often combined with antibody-staining techniques to highlight aspects of the muscle fibers. For example, antibodies to mitochondria-specific proteins can attach to those proteins and highlight them under a microscope. This makes it much easier to pick out those structures. Antibodies are also the primary way of telling the difference between fast- and slow-twitch muscles. Antibodies to fast myosin highlight the fast-twitch fibers, whereas antibodies to slow myosin highlight slow-twitch muscle fibers. Under a microscope, it becomes easy to see the ratio of fast- to slow-twitch fibers in the different biopsy samples.

Experiments using biopsy samples are also how researchers learned what types of chemicals are present in fast- and slow-twitch muscles. To look at levels of chemicals in the muscle, researchers usually study an athlete on a treadmill doing a particular type of exercise. At specific times during that exercise, the researcher takes a biopsy sample and grinds it up to release the chemicals into a solution. Researchers can then look at levels of ATP, creatine phosphate, calcium carbonate, lactate, or the different enzymes involved in making ATP.

With these types of experiments, people now know how quickly muscles use up creatine phosphate, when they begin producing lactate, and what the conditions are like within the muscles when an athlete begins feeling burning pain or fatigue. Biopsies are particularly powerful when the researcher can both look at the physical aspects of the muscle and compare it with information about what is happening within the muscle at the same point in time.

Although biopsy experiments have given a good overall picture of how muscles behave during exercise, there are still many unanswered questions. Some of these have to do with how the muscles regulate the switch between slow- and fast-twitch fibers, how they control the fuel they use for energy during exercise, and what specific factors contribute to fatigue. These questions are particularly important to athletes such as those in the Olympics or in professional sports such as football or cycling. Learning what triggers muscles to be more efficient or fatigue slower could give savvy athletes an edge over their competitors.

One way to address these questions is to use gene chip experiments that reveal what genes are being used in a muscle at a given point in time. Although the questions being asked have to do with how a person reacts to exercise, the answers may come from learning what genes individual muscle fibers use. By comparing resting muscle, exercising muscle, and recovering muscle, researchers may start to learn what genes are being used—and therefore what proteins are present in the muscle during that phase of exercise. Eventually, this information could help trainers learn exactly what is happening in each athlete's muscles and customize the training regimen for particular athletes.

Muscle Metabolism

Muscle biopsies give a detailed glimpse of the events taking place in a tiny subsection of a muscle. In some cases, however, researchers may want to know how the whole body is responding to exercise, whether the muscles are burning more fat or more glycogen, what the VO_2max or lactate threshold is for the athlete, or how the muscles are generating ATP. Answering these questions generally involves putting an athlete on a stationary bike or treadmill and measuring his or her breathing and heart rate. In many cases, this type of experiment will include muscle biopsies taken at specific points in time to correlate the events taking place in the whole body with events in the individual muscles.

To measure VO_2max, the athlete generally wears a breathing tube attached to a computer. A machine reads the amount of oxygen in the air being breathed and the air being breathed out. From that, a computer subtracts the oxygen an athlete actually used to create ATP. At a certain intensity of exercise, the athlete stops being able to use more oxygen despite the fact that the exercise intensity increases. This point is the VO_2max.

Although these tools have been standard for many years, they are still a regular part of research in exercise physiology. As with most research in this field, the end goal is generally to find the best way to train professional athletes. By testing athletes on a treadmill throughout their training, researchers can learn more about the best way to train athletes for different events or conditions. Coaches also use this information to modify an athlete's training. Particularly in the case of endurance athletes such as runners, cyclists, or cross-country skiers, coaches rely on information from exercise physiologists to make decisions about the best training strategies.

7

Muscular Injuries

On a day-to-day basis, most people don't think much about their muscles. It's not until the muscles hurt due to a **strain** or damage from a challenging workout that they make their presence known. At that time, it becomes clear just how often a person uses a particular damaged muscle.

What is surprising is that pain is a poor indicator of how much damage a muscle has sustained. Some injuries don't cause pain, and in others the pain disappears long before the injury has healed. This discrepancy can lead people to go back to exercising before a muscle is ready—and further injure the muscle.

Most muscle injuries are due to contracting the muscle as the muscle extends, also called eccentric exercise. In a normal workout, eccentric exercise can damage individual sarcomeres, making the muscle hurt and become weaker for several days. This type of muscle pain is common in people who are just starting to exercise or in well-trained athletes who increase or change their workout.

Common muscle pain due to a workout goes away after a few days and doesn't cause any long-term damage to the muscle. A more severe injury to the muscle, called a muscle strain, can leave permanent scar tissue in the muscle and can take as long as several months to heal. This type of injury happens when the muscle is pulled too far, tearing the muscle tissue.

DELAYED ONSET MUSCLE DAMAGE

Assessing the Damage

PAIN

Pain is a poor indicator of muscle damage, but it is certainly one of the most noticeable. Chapter 3 described two kinds of muscle pain: acute and delayed. Acute muscle soreness comes from fatigue during a workout. Exercise physiologists aren't sure precisely what causes this pain. It may have to do with lactic acid building up in the muscle or with cramping when the muscles aren't getting adequate chemicals from the blood to keep them contracting normally. Sports beverages that contain calcium and salts, which the muscle uses in the process of contracting, can help prevent pain due to cramps.

Delayed onset muscle soreness (often referred to as DOMS) is different from the burning pain felt during exercise. A person may end up with DOMS after a workout that wasn't painful or may have a painful workout but no DOMS. This type of pain usually starts the day after the workout and is particularly bad after a workout that includes eccentric action such as running downhill. Often this pain is worst near the muscle's insertion, or the end farthest from the center of the body. For example, the thigh muscle will hurt worst near the knee, and the biceps will hurt worst near the elbow. DOMS usually goes away within a few days, depending on how difficult the workout was. Figure 7.1 shows how long it takes a muscle to recover from aspects of muscle damage due to exercise.

DYSFUNCTION

After a particularly hard workout, the muscles may become weaker for several days. Any exercise that causes a loss of strength probably also results in DOMS, but even after the pain has gone away the muscle may still have enough damage that it is weaker than normal. After some particularly damaging exercise, muscle power can go down by as much as half. This type of dramatic reduction in strength usually becomes obvious the day after damaging exercise and recovers over the course of a week to ten days.

PHYSICAL DAMAGE

Several types of damage within the muscle may cause the muscle to lose strength. It could be that the sarcomere itself is damaged and can't contract normally. This is the type of damage that usually happens after eccentric exercise. In severe cases, the sarcomeres may be stretched so much that the actin and myosin no longer overlap—a process also known as popping. These sarcomeres can no longer contract, making the whole muscle weaker.

Along with the sarcomeres popping, the Z disks—the dark bands of pro-

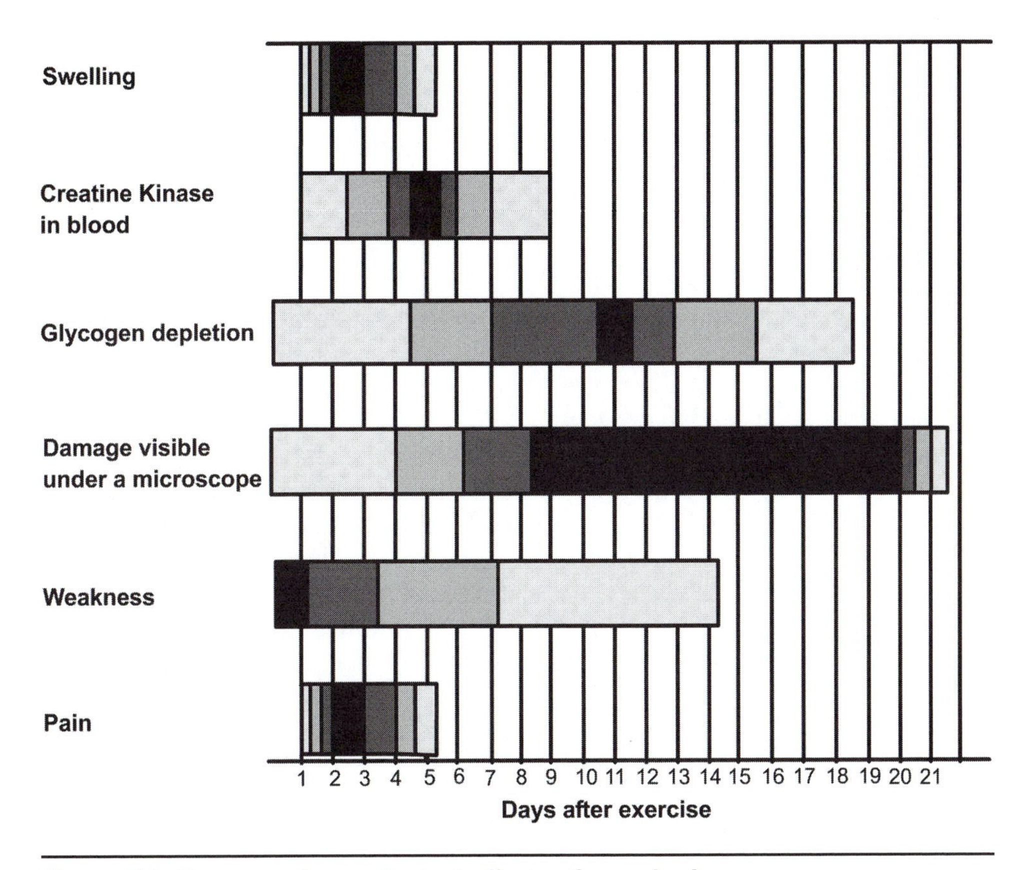

Figure 7.1. Recovery from physical effects of muscle damage.

tein that separate the sarcomeres from each other—pull apart in a process known as streaming. Without an intact Z disk for the actin filaments to embed into, the actin filaments can't exert any force to shorten the sarcomere.

Damage from eccentric action also damages the structural proteins that hold the sarcomere together. Some studies have found these proteins disrupted for as long as two weeks after the initial muscle damage. Although they aren't absolutely necessary for contraction, disrupting these proteins does prevent the sarcomere from contracting with its full strength.

In some types of muscle damage, it may be the sarcoplasmic reticulum that is injured. In this case, the muscle itself may be capable of contraction, but the signal to contract doesn't reach the muscle because the sarcoplasmic reticulum doesn't transmit a signal normally.

After a particularly hard workout, such as running a marathon or a long bike ride, the muscles look very different under an electron microscope than normal healthy muscle. The sarcomeres are stretched apart, the Z disks do not form solid bands, and muscle fiber membranes have gaps or ruptures. These factors all add together to make the muscle weaker until the damage heals.

CHEMICAL CHANGES

Some muscle damage is purely physical—the filaments in the sarcomere no longer overlap and can't contract. But other damage is more indirect. As a muscle becomes extremely tired, it no longer makes enough ATP. Not only does that make it hard for the muscle to contract, but it also means that the muscle can't carry out normal functions.

One example of this is that the muscle uses ATP to pump calcium back into the sarcoplasmic reticulum after a signal to contract. Without ATP, calcium builds up in the fiber and in the mitochondria. It turns out that although calcium is essential for many normal functions, it can also be toxic to the muscle in high doses and can completely prevent the mitochondria from making ATP. Large amounts of calcium can cause damage to the DNA, the muscle membrane, and the structures within the muscle. In the end, making too little ATP can lead to long-term physical damage in addition to making the muscles feel tired.

Biological Indicators of Muscle Damage

Chapter 6 described how researchers remove small samples of a muscle using a biopsy needle. Muscle biopsies can reveal muscle damage on a very small scale, but they are painful for the patient and many techniques for examining the biopsy are time consuming, expensive, or require specialized equipment such as an electron microscope. However, with the proper equipment a biopsy sample can give a doctor a lot of information about whether a muscle is structurally normal and contains all of the proper chemicals.

In addition to physical indications of damage, in some types of muscle damage proteins from the muscle will leak out into the blood where they can be measured. One such protein is an enzyme called creatine kinase. This enzyme adds a phosphate to creatine to generate the energy source creatine phosphate. Normally, the blood does not contain any creatine kinase. If it is present in a blood sample, that means that the muscle membrane has been damaged and the enzyme has leaked out of the muscle into the blood. This type of analysis is often used to assess damage to the heart muscle after a heart attack.

Another protein that can seep into the blood after damaging exercise is myoglobin. This protein takes up oxygen from the blood and is particularly common in slow-twitch muscles that rely on a steady amount of oxygen to function normally. Myoglobin appears in the blood immediately after damaging exercise such as long-distance running, whereas creatine kinase can take a full day before it shows up in the blood. In fact, after extremely long-distance runs, myoglobin can be as much as twenty times more concentrated in the blood.

Creatine kinase and myoglobin leak into the blood over the course of several days, with the most protein seen in the blood at about four days after exercise (for the difference between men and women, see "Sex Differences in Creatine Phosphate Release"). This delay isn't easily explained. In people who have surgery that involves cutting the muscles, small amounts of the same proteins leak out of the muscles into the blood, but they reach peak levels at about two days after the surgery, then subside. The question is why it takes longer for proteins to leave the muscles after exercise than after direct injury such as being cut in surgery.

One theory has been that the muscle tries to repair initial damage to the sarcomeres. Over time, however, the damage is too great and cannot all be repaired, so some of the muscle fibers break down. These fibers are what might eventually release creatine kinase into the blood. This theory breaks down when researchers realize that, in other experiments, the amount of sarcomere damage does not always correlate with how much creatine ki-

Sex Differences in Creatine Phosphate Release

Men and women release different amounts of creatine kinase into the blood after muscle-damaging exercise. Men release a much larger amount than women. Some researchers wondered if this difference had to do with the types of hormones that men and women make, because they have the same levels of creatine kinase in the muscles. When Tiidus and Bombardier gave male rats the hormone estradiol—which is at much higher levels in women than in men—and made the rats run on a treadmill, the rats released much less creatine kinase than their normal male peers. The males that were given estradiol also had less muscle damage when researchers looked at a muscle biopsy under a microscope. By removing the ovaries of female rats and thereby lowering estradiol levels, the rats produced the same amount of creatine kinase as males after exercise. Other studies on isolated muscles confirmed that estrogens, such as estradiol, protect the muscle membrane and prevent proteins such as creatine kinase from leaking out into the blood.

This study makes clear that men and women respond differently to exercise. Many athletic studies are done on elite male athletes, but new studies such as the ones looking at muscle damage in male and female rats are causing sports medicine doctors and coaches to take into account a person's gender when devising a workout strategy or when assessing an athlete's injuries.

nase is released into the blood. In another set of experiments, an athlete had three extremely difficult workouts spread out by two-week intervals. After each workout, the athletes had lost as much as 50 percent of their muscle strength due to muscle damage. Creatine kinase levels in the blood increased dramatically after the first workout, but didn't go up after the next two workouts. This told the researchers that muscle damage itself isn't enough to release creatine kinase into the blood.

Preventing Muscle Damage

If hard exercise causes muscle damage, it's no surprise that so many people would be interested in preventing that damage from occurring. Professional athletes have a vested interest in not losing training days, but even casual athletes don't want to miss out on a favorite form of exercise because of pain or muscle weakness.

TRAINING

Regular training is the best way to prevent damage, even from eccentric exercise such as downhill running. In fact, several studies have shown that a single strenuous bout of exercise can protect against damage for several months. How exercise protects against further damage is not entirely known. After regular exercise, the muscles become stronger, have more exercise-specific enzymes, have better blood supply, and have more fuels available for energy. But even a single stint of eccentric exercise can protect against further damage.

Several changes may take place in the muscle to protect against muscle damage after eccentric exercise. One idea is that after damaging eccentric exercise, the nervous system simply learns not to allow the muscles to contract to the point of damage. The nervous system may recruit different motor units, or a great number of motor units, for the same type of exercise to prevent further injury. Other researchers have noticed that individual sarcomeres become shorter after eccentric exercise. These shorter sarcomeres are more resistant to being pulled apart.

Although nobody has proven any of these theories, some researchers have found that a few days after a second bout of damaging eccentric exercise, less creatine kinase was present in the blood than after the first attempt at the same exercise. This seems to indicate that the muscle became less damaged, possibly because of changes in how the muscle was used.

STRETCHING AND MASSAGE

Many athletes stretch their muscles before exercise in the hopes of preventing muscle damage. If damage does occur, then massage certainly feels good. However, studies vary in terms of whether these measures actually help prevent or heal the damage. Some studies have found that people who stretch have less pain after a workout than people who do not stretch. However, Johansson found no difference in muscle pain between stretched and unstretched legs in women cyclists.

Part of the problem in understanding whether stretching prevents DOMS is that it is hard to regulate how much people stretch and how hard they work out, and each person feels pain differently. With these challenges, it has been difficult to find out if stretching or massage actually helps prevent muscle pain or muscle damage. Studies have found that on the whole, people who do stretch have as much pain, creatine kinase in the blood, and weakness in the days following exercise as people who do not stretch or get a massage after the exercise.

VITAMIN E

One idea has been that when muscles are damaged, highly reactive molecules called free radicals are released into the muscles. These free radicals can cause damage to the cell, cellular components, and DNA. Antioxidants such as vitamin E can protect against free radical damage. If muscle damage were due to free radicals, then taking vitamin E regularly may be able to protect against the damage. Another argument that free radicals might be involved in exercise-related muscle damage is that trained athletes have high levels of natural antioxidants to soak up free radicals and protect their muscles from damage.

Some studies in rats have shown that when vitamin E is in short supply, more creatine kinase builds up in the blood after running. However, giving the mice vitamin E did not prevent muscle damage due to eccentric exercise. Exactly how vitamin E may protect against muscle damage, and whether it protects against all types of damage, is still an ongoing debate.

HORMONES

Men and women differ in terms of how damaged their muscles become after strenuous exercise. Researchers think this difference may be because estrogen, which is at higher levels in women than in men, protects against free radicals similar to the way vitamin E does. In keeping with this, Amelink and colleagues found that male rats with a shortage of vitamin E are much more prone to muscle damage than female rats with too little vitamin E. The thought is that the estrogen acts as a backup for vitamin E in the female rats.

Although giving male rats estrogen does make them less prone to muscle damage, this is hardly the solution most male athletes would prefer. Estrogen makes men more like women in a number of ways, which detracts from the benefit of having more resilient muscles. An artificial form of estrogen called tamoxifen mimics some aspects of estrogen but not all. It protects male rats against muscle damage and doesn't make the rats more feminine. This drug, or other estrogen mimics, may be one way to protect men's muscles.

Repairing Muscle Damage

NATURAL REPAIR MECHANISMS

Mature muscle cells are not able to divide and make up for fibers that are so damaged they can no longer function. Instead, repair is left to a group of immature cells called satellite cells that dot the outside of muscle fibers. As discussed in Chapter 4, these cells have begun down the developmental pathway to become muscle cells, but they stall before they mature. They are committed to becoming muscle cells but haven't yet finished developing.

Researchers do not know all the signals from the damaged muscle that nudge the satellite cells to resume their division. Whatever those signals may be, nearby satellite cells divide several times, leaving one cell behind as a satellite cell while the remaining new cells fuse with the damaged muscle and with each other to bolster the damaged fibers. This process can cause the individual fibers to grow larger, which is one of the ways that the muscles become larger after exercise.

It turns out that muscles in young people heal more quickly than muscles in older people. In a study using rats, the young rats showed quite a bit of damage immediately after two hours of running on a treadmill, but that damage all but disappeared within a week of the exercise. In adult rats, the researchers saw less damage at any given time, but that low level of damage lasted for as long as two weeks.

In the same experiment, the young rats had many activated satellite cells within hours after the exercise. In adult rats the satellite cells didn't become active until days or weeks after the damaging exercise. Although the researchers haven't done this exact experiment in humans, it does give some clue as to why people seem to heal more slowly as they age.

Many of the same factors that control muscle development in the embryo (see Chapter 4) also seem to control when satellite cells resume their own development. Some of these growth factors are bound to the outside of the muscle fiber and are released when the fiber is damaged. Others are made by immune cells that clean up debris from the damaged muscle.

NONSTEROIDAL ANTI-INFLAMMATORY DRUGS

Because muscle damage often hurts, many athletes take painkillers such as ibuprofen in the days following extreme exercise. Ibuprofen is in a class of drugs called **nonsteroidal anti-inflammatory drugs (NSAIDs)**. As the name indicates, these drugs are not steroids and can reduce inflammation. In addition to fighting pain, many people thought these drugs might also help speed healing. When muscle fibers are damaged, immune cells enter the muscle to clean up the damage. Having immune cells in the muscle can cause swelling and make the initial damage even worse. Knowing that, it is logical to think that by reducing swelling the injury may heal faster.

When researchers gave NSAIDs to animals after rigorous exercise, the animals gained strength much faster than their counterparts who did not receive the painkillers. Initially, this result looked as if NSAIDs could be a boon to injured athletes. However, by one month after the injury the undrugged animals had recovered more strength than those taking painkillers. From this experiment, it seems that drugs such as ibuprofen may feel good to an injured athlete, but may slow healing in the long run.

MUSCLE STRAINS

A muscle strain, or "pull," is the most common severe muscle injury in athletes. As the name implies, a muscle strain happens when the muscle is pulled with too much force, tearing the muscle tissue. Unlike DOMS, which causes a small amount of damage throughout the muscle, strains usually rip fibers in a small area of the muscle. They also tend to hurt immediately rather than taking a few days for the pain to buildup.

Strains tend to happen when a muscle is shortening such as by lifting a barbell with the bicep rather than during eccentric exercise. The muscle is strongest when shortening and can pull hard enough to physically rip the muscle fibers.

A pulled muscle is the mildest version of a muscle strain and doesn't cause lasting physical damage to the muscle. A partial tear happens when the muscle tissue tears and the two sides of the tear contract and pull apart, leaving a small indent in the muscle. This indent fills with blood and eventually swells and is warm to the touch. The torn section is slowly replaced by scar tissue. A complete tear or rupture is the most severe form of a strain. In a rupture, the muscle completely or almost completely tears apart. This kind of damage often requires surgery for the muscle to heal normally.

Commonly Strained Muscles

The muscles that are most prone to strains are ones that cross two muscle joints. For example, the biceps muscle is attached at one end to the shoulder, crosses the shoulder joint, then crosses the elbow joint before attaching just below the elbow. This muscle spans two joints and therefore has the most likelihood of getting stretched excessively. Other muscles that are prone to strain are the ones that are mainly used to limit a joint's motion. The hamstring running down the back of the thigh, for example, is generally used to limit how far the knee can extend when the hip is flexed. If the hamstring attempts to contract and limit that motion while the knee keeps extending, that can overextend the hamstring and cause a muscle strain.

Sprint running is almost completely an eccentric action, which is why sprinters are prone to delayed muscle soreness. The amount of force involved in sprinting can also make these athletes prone to muscle strains. Strains are also common in sports where a person changes direction or speed quickly such as soccer, football, and rugby. The strains are likely to happen in muscles with a lot of fast-twitch muscles, but that could be because the types of activities that lead to strains also cause an athlete to have more fast-twitch muscles rather than because those muscles are specifically more prone to strains. It could also be that fast-twitch muscles contract with more force, making them strong enough to pull the muscle apart.

CALF

Strains in the calf muscle are most likely to happen when a person is on the toes with the knee bent. In that position, the muscle is contracted tightly to hold the heel off the ground. If an athlete bounces so that the heel hits the ground while the calf muscle is still contracted, that could stretch the muscle enough to cause a strain. This type of injury is so common in tennis that it is called "tennis leg." It is also common in track and field events that involve sprinting, as well as in boxing and gymnastics.

In calf muscle strains, the person knows immediately that he or she has had a severe injury. The calf feels a sharp pain as if the muscle had been hit with a stone, and it often makes a loud cracking noise. The person usually can't use the muscle after the injury. Within a few days, a bruise usually shows up at the lower end of the muscle, near the Achilles tendon, although the sorest part of the muscle is usually a bit higher in the muscle.

QUADRICEPS

The quadriceps are a group of four muscles running down the front of the thigh. They all work together to bend the hip and pull up the knee toward the stomach. These muscles are most likely to tear in sports such as football, ice hockey, skiing, and track and field events that involve sprinting. The more severe forms of strains in the quadriceps muscles can leave a dent in the front of the thigh where the muscle pulls apart. People use the quadriceps all the time as they walk around or go up stairs. Because the muscle is used so much, it is very prone to getting reinjured after the initial strain.

BACK

Back strains happen when the back muscles are contracted, then pulled by a sudden twist or stretch in the back. Some common ways to strain back muscles have nothing to do with exercise. For example, sneezing while reaching out an arm can suddenly pull on the back muscles that were supporting the arm, causing a strain. Back strains are also common in rowing and skiing.

The pain from a back strain is usually completely disabling, because the back muscles are used all the time to hold a person upright. Simply sitting or standing can be excruciating. Keeping the back muscles strong through regular exercise can help support the back and prevent injury.

BICEPS

Strains in the biceps muscle happen when the biceps is contracting to hold the elbow joint closed while the elbow is pulled straight. This commonly happens in pole vaulting, discus throwing, and boxing. In almost all cases the injury happens in the lower third of the muscle, near the elbow. For several days after the injury, it can be extremely painful to straighten the elbow. These injuries usually heal completely within a few weeks.

Location of Muscle Strains

Strains generally damage the region of a muscle where it connects to a tendon, but they can also take place in the junction between the tendon and the bone, in which case the muscle is held to the bone by only a portion of the tendon. In the junction between the muscle and tendon, the muscle fibers and tendon are intertwined in a strong meshwork that can take up as much as half the length of the muscle in some cases. In the hamstring, for example, the tendon that attaches the muscle to below the knee extends almost the whole way along the back portion of the muscle. In the hamstring, a person can have a muscle strain at almost any point in the muscle and still have that strain occur at the junction of the muscle and the tendon. Figure 7.2 shows the extensive junction between tendon and muscle in the hamstring.

When researchers mimic muscle strains in animals, they have found that whether they cause the strain by pulling on the muscle when it is contracting or when it is relaxed, the strain is still most likely to happen at the junction of the tendon and muscle, and usually happens at the end of the muscle that is farthest from the center of the body.

The connection between muscle and tendon tissues appears to be strong, so researchers do not understand why that region is the weakest point and the first to pull apart under strain. In one set of experiments, researchers took a rabbit muscle and painted small black dots along its length. They then videotaped the muscle as they quickly stretched it to mimic a strain. What they saw on the video is that the dots move farther apart as the muscle stretches but were still spaced an equal distance apart throughout the bulk of the muscle. At the junction between the muscle and the tendon, however, the dots were stretched farther apart than at other points, indicating that this point in the muscle is where most stretch occurred. The researchers assume that this is why that region gets damaged first by too much stress.

In a strain, muscle tissue physically pulls away from the tendon, so any muscle contraction doesn't translate to the tendon and pull the bone. The

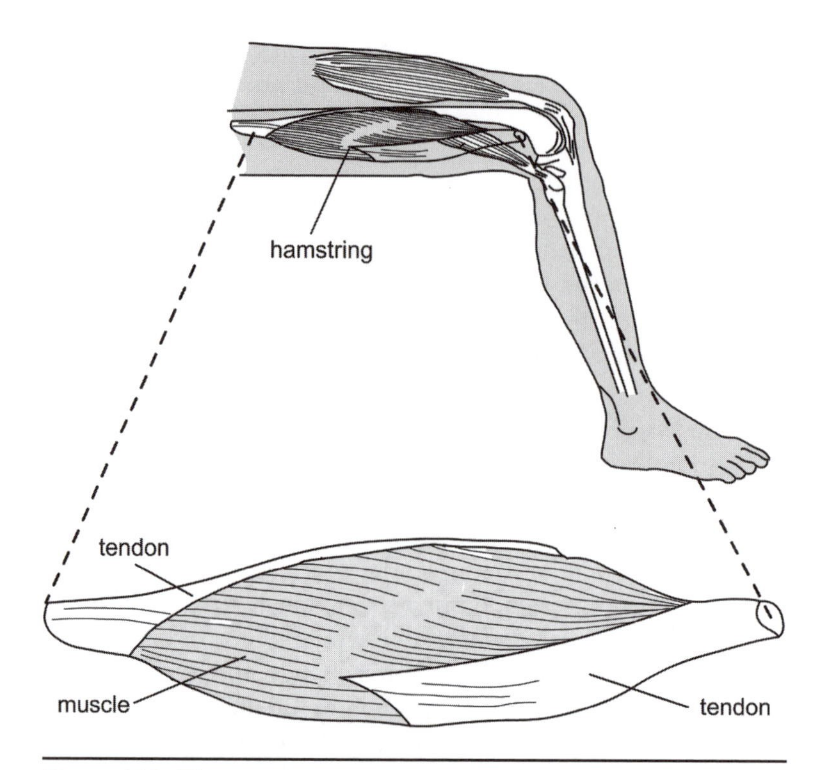

Figure 7.2. The tendon-muscle connection in the hamstring.

severity of the strain depends on how much the muscle has pulled away from the tendon—if it is just a small percentage of tissue, then the injury is relatively mild and will probably heal normally. In a severe strain, as much as 90 percent of the tissue can be pulled away.

Treatment for Muscle Strains

The goal of any treatment for a strained muscle is to have the scar tissue that builds up be soft and flexible, and to have the muscle regain its original strength. Treating a muscle strain usually involves resting the muscle, applying ice to reduce the swelling, and wrapping the muscle to immobilize it and prevent it from contracting to cause further damage. After the injury has healed somewhat, physical therapy can help get strength back in the muscle. Some doctors also prescribe drugs to help a person deal with the pain, or muscle relaxants to prevent the muscle from contracting and causing further damage. If the tendon has actually pulled away from the bone, surgery may sometimes be necessary to make sure the tendon reattaches properly.

Taking drugs such as NSAIDs can help a person with the severe pain from a muscle strain and also reduces swelling, but doctors usually only recommend those drugs for a few days immediately after the injury. Taking the drugs after that could prevent the muscle from healing properly.

Whatever treatment strategy a doctor recommends, the healing process is the same as with any other injury. Satellite cells respond to the damage by dividing and fusing with damaged muscle fibers. In the region where the muscle fibers were ripped, scar tissue builds up to hold the two sides of the tear together.

In the past, doctors had thought that keeping the muscle immobile until it was fully healed was the best way to recover from a muscle strain. Now doctors are more likely to tell an injured athlete to begin moving early in the healing process. This is because when a muscle is not used for a long time, it loses strength and the muscle fibers shrink. Weaker muscles are less able to resist further damage, so immobilizing the muscle actually makes that muscle more prone to injury once the athlete starts using it again. By using the muscle in a limited fashion within three to five days after the injury, the muscle remains strong enough to resist further damage. Moving a muscle also seems to make the muscle tear heal with a smaller, more flexible scar that interferes less with the healed muscle.

People who have strained muscles should be extremely careful about when they begin getting muscle massages. A massage before the muscle has healed completely can break up the scar tissue and cause even more damage to the muscle fibers. If the fiber is damaged repeatedly, bone can start to form within the muscle itself. This process is called muscle ossification.

Having bone growing within the muscle can be extremely painful and interferes with nearby joints. It also makes the muscle less flexible and more prone to another injury. Once the bone has finished forming, which can take as long as two years, a doctor can remove the bone during surgery. This surgery then heals like a normal tear as long as the person avoids massage or further damage. If the surgery takes place while the bone formation is still taking place, then the person could end up with even more bone building up than before the surgery.

Prevention

The only way to prevent muscle strains is by being careful to avoid situations that may lead to the injuries. Some athletes stretch before a workout and warm up slowly at the beginning of each workout to give the muscles a chance to loosen up before they are used rigorously.

There are no good, large-scale studies that have looked at whether stretching or warming up actually prevent muscle strains. William Garrett has found that exercise regimens that include both of these measures do tend to produce fewer injuries. However, people who follow those precautions are also more likely not to overexert their muscles or put themselves in positions where a strain would be likely. From this study, it isn't clear that stretching and warming up alone are enough to prevent injury. An exercise regimen that includes these activities in addition to being careful about overexertion seems to be the only effective way to prevent muscle strains.

STRETCHING

Regular stretching makes a muscle more able to contract even when it is stretched farther than is comfortable. Because muscle strains happen when a muscle is pulled beyond its normal limit, it seems reasonable that stretching could help a muscle resist strains.

In animal studies, when researches stretched a muscle to a certain length, the muscle became significantly less tense with each stretch. The differences in how tight the muscle became were most noticeable with the first four stretches. If the muscle is less tense at a longer length, then that muscle would also be less likely to tear at the longer length.

One question has been what the actual effects are of stretching the muscle. The muscle fibers themselves don't appear to change after stretching. Instead, some researchers think that it is the tendons that stretch, along with proteins between the muscle fibers that hold the muscle in place. Stretching may loosen up this mesh of connective proteins and allow the muscle to move normally.

WARMING UP

In laboratory studies where researchers make animal muscles contract slowly several times before having them lift heavy loads, those muscles had a stronger contraction when stretched than muscles that weren't warmed up. It also takes more force to strain muscles that have warmed up versus the muscles that have not warmed up. As the name implies, these muscles also got warmer during the slow, easy exercise. When researchers artificially warm or cool muscles, they find that cold muscles become damaged when stretched only a small amount, whereas warm muscles can stretch much farther before the muscle pulls away from the tendon.

These studies of stretching and warming up were done in animals in a laboratory, so they don't apply directly to human athletes, but they do suggest that warming up and stretching are good ideas for people who want to prevent injuries. They also suggest that athletes who are exercising in cold conditions should be careful to start their workout slowly to give their muscles a chance to get warm.

MUSCLE CUTS OR BRUISES

Muscle cuts and bruises are not the type of exercise-related injuries that most casual athletes expect to encounter, but they can be common, especially in impact sports like football or rugby.

Cuts

Any cut that goes through the skin and fatty layers beneath the skin runs the risk of slicing muscle. If it is just a small cut, the muscle will still have enough intact fibers to still contract normally, though it will likely be painful. But larger cuts that sever a significant portion of the muscle can prevent that muscle from contracting.

As with a severe strain, in which the muscle has pulled away from the tendon, a muscle cut begins healing with stiff scar tissue. This scar tissue will hold the two sides of the muscle together but does not become part of the muscle and is never able to contract. The goal of any treatment is to have that scar tissue be as soft as possible so it does not physically make it even harder for the muscle to contract.

Each muscle fiber spans the entire length of the muscle and has only one nerve connection. Any cut that goes through a fiber will leave one half of the fiber with a nerve connection, while the other half is severed from the nerve. Scar tissue eventually builds up between these two halves, which never rejoin. Even after the cut has healed and the scar is fully formed, only the portion of that fiber with a nerve connection can ever contract again. Depending on how deep the cut was, this could leave a person with a permanently much weaker muscle. In some animal studies, cuts that go all the way through a muscle can heal, but the muscle can lose as much as half of its original strength.

Bruises

Bruises to the muscle go beyond the purple patch that results from bumping into a chair. These day-to-day bruises only damage tissue just under the skin and leave the muscle unharmed. The muscles usually only get bruised in high-impact sports where two athletes run into each other at high speeds—especially if the athletes have on hard padding or helmets.

Although a bruise does not sound like a serious injury, it can take a long time to heal from a bruise to the muscle. As with any other types of muscle damage, the fibers that are directly impacted release proteins into the blood and aren't able to contract properly. The area that got the bruise usually swells with blood that leaks out of damaged blood vessels, then with immune cells that clean up the damaged fibers. Eventually, the blood goes away and the muscle begins to form scar tissue.

As with other types of muscle damage, the scar tissue that forms is not ever able to contract. In recent years, sports medicine doctors have been recommending that people with muscle bruises begin moving as soon as possible to help the muscle stay strong. Using the injured muscle also helps form a much smaller scar that interferes less with the muscle.

Although muscle bruises do form scars, they are a much less serious injury and usually heal both faster and with less long-term damage than a strain. The difference is in the way the muscle fibers are damaged. In a muscle strain—or muscle cut—the injury goes across muscle fibers, so many fibers get severed. With a bruise, the fibers on the outside of the muscle have the most damage while the fibers deeper inside the muscle are protected. It is only these outer fibers that usually end up with scar tissue.

MUSCLE KNOTS

Muscle knots are hardly the most serious muscle-related injury, but they can be annoying and for some people intensely painful. Muscle knots, or nodules, are hard lumps that form in the muscles. The lump comes from a small region of the muscle where fibers are chronically contracted. Under a microscope, the knot region has tightly contracted sarcomeres. Farther along the fiber, the sarcomeres are stretched out to make up for the knotted region.

Doctors are not sure why small parts of the fiber contract into a knot. It seems to happen most often in muscles that are used regularly but at low intensity, such as the muscles of the back that hold a person upright. They are particularly common in the backs and shoulders of people who spend a lot of time at a computer, or in muscles that are used regularly for a particular type of exercise. For any muscle, the knots are most likely to form at particular locations called trigger points.

Although anybody can get muscle knots, they tend to be more common in people who smoke, are overweight, have diabetes, haven't been sleeping properly, or are deficient in some vitamins. They can also be an early sign of muscle stress in athletes. If a muscle is being used regularly and forms a knot, that could be a sign that the athlete isn't using the muscle properly or that the muscle is not strong enough for that exercise load. Finding the knots and making appropriate changes in training can prevent a more serious muscle strain in the future.

The best treatment for muscle knots is massage. Putting pressure on the contracted region can force those sarcomeres to relax, eliminating the lump. Once a knot has formed it is likely to come back in the same spot, in part because the muscle will probably be under the same stress that caused the initial knot. Sometimes repeated massages along with changing posture, office setup, or athletic habits can prevent the knot from coming back. Regularly stretching the muscle can also prevent a knot from forming in the first place.

TENDON STRAINS

Sports that require the same movement over and over, such as tennis, baseball, or swimming, can damage the tendons that attach muscles to the bones. Although these injuries do not harm the muscles themselves, they are due to repeated muscle use, and some forms of rehabilitation involve building muscle strength in order to protect the tendon.

Tendons are made up of a tough, fibrous protein called **collagen**. The long collagen fibers intertwine to form an elastic meshwork similar to a strong rope. When muscles pull on the same tendon repeatedly, the collagen can begin to pull apart, forming tiny tears called microtears.

Tendon strains are very much like muscle strains—the tendon tissue begins to tear when the tendon is overstretched. As with muscle strains, tendon strains can be relatively mild with only a small amount of **microtears**, or they can be relatively severe. In the most severe cases, the tendon tissue is almost completely pulled apart.

When the tendon has microtears, immune cells from the bloodstream enter the tendon and clear away the damaged tissue. The swelling from these immune cells can cause even more problems. The larger tendon can begin rubbing on surrounding tendons, ligaments, or bones, which can make the tendon itself more painful or can damage those surrounding structures. Icing the tendon keeps this swelling down and can speed up how quickly the tendon heals.

Tendons can heal normally if they are allowed to rest and if the injury was relatively minor. With more severe injuries, however, scar tissue builds up and the tendon is never as flexible as it was originally. Being less able to stretch, that tendon is then more likely to be strained in the future. To prevent this cycle of tendon injury, sports medicine doctors often recommend stretching the muscle and tendon regularly. That way, although the tendon itself is not as flexible, the overall tendon and muscle can stretch farther and not be injured by stress.

LIGAMENT SPRAINS

Any strain refers to damage to a muscle or tendon, whereas a **sprain** refers to damage to the ligaments. Ligaments are made of the same rope-like collagen as tendons, but they support joints rather than connecting muscles to bones. As with tendon strains, ligament sprains do not damage the muscle itself but can happen when the muscle is not strong enough to support the joint.

By far the most common joint to sprain is the ankle. Any time the foot rolls in an unusual position, it can stretch the ligaments and cause painful microtears. In mild sprains the ligaments swell and are painful, but they can still support the ankle joint. In more severe sprains, the ligaments are so torn that they can't support the ankle. In these cases, the ankle needs to be wrapped to hold the joint in place and prevent more tearing to the ligaments.

As with tendon strains, the best treatment for sprains is ice. That keeps the ligaments from swelling and getting even more damaged. It can also take some immediate pressure off of the ligaments and make the injury much less painful. Sports medicine doctors often recommend that people who are prone to ankle sprains strengthen muscles around the feet, calves, and shins to help support the ankle. If these muscles are strong, then the ankle is less likely to roll and pull on the ligaments.

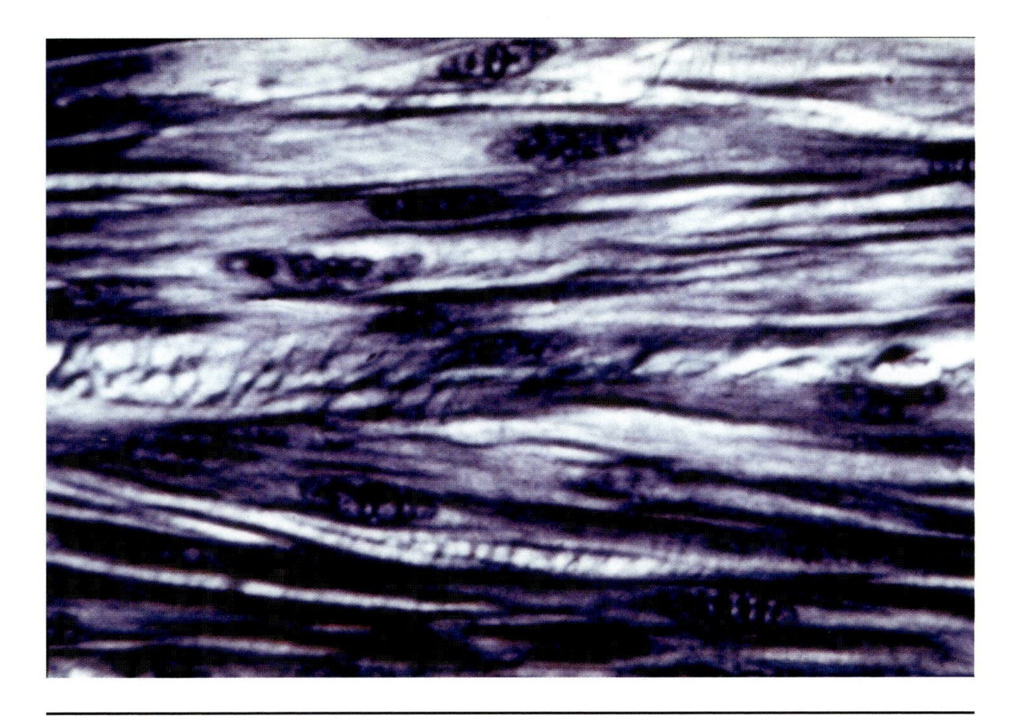

Smooth muscle. © Educational Images/Custom Medical Stock Photo.

Cardiac muscle. © Fred Hossler/Visuals Unlimited.

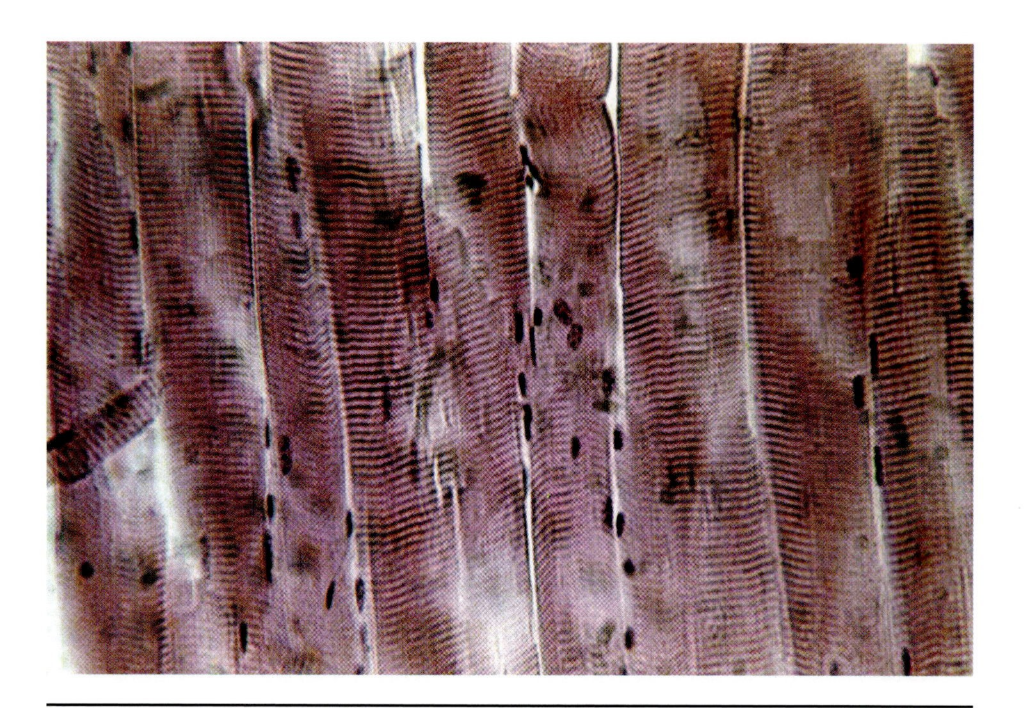

Skeletal muscle. © Educational Images/Custom Medical Stock Photo.

Skeletal muscle showing sarcomeres. © Custom Medical Stock Photo.

Aerobic exercise. © Hulton/Archive.

Anaerobic exercise. © Skjold Photography.

Stretching before exercise. © Hulton/Archive.

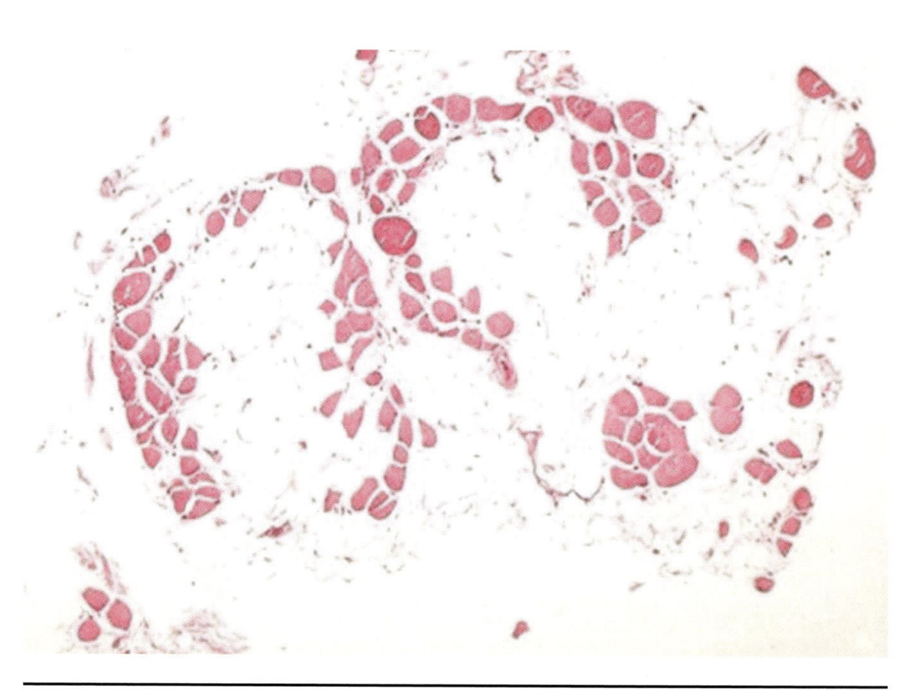

Muscle biopsy showing muscle replaced with fatty tissue. © Centers for Disease Control.

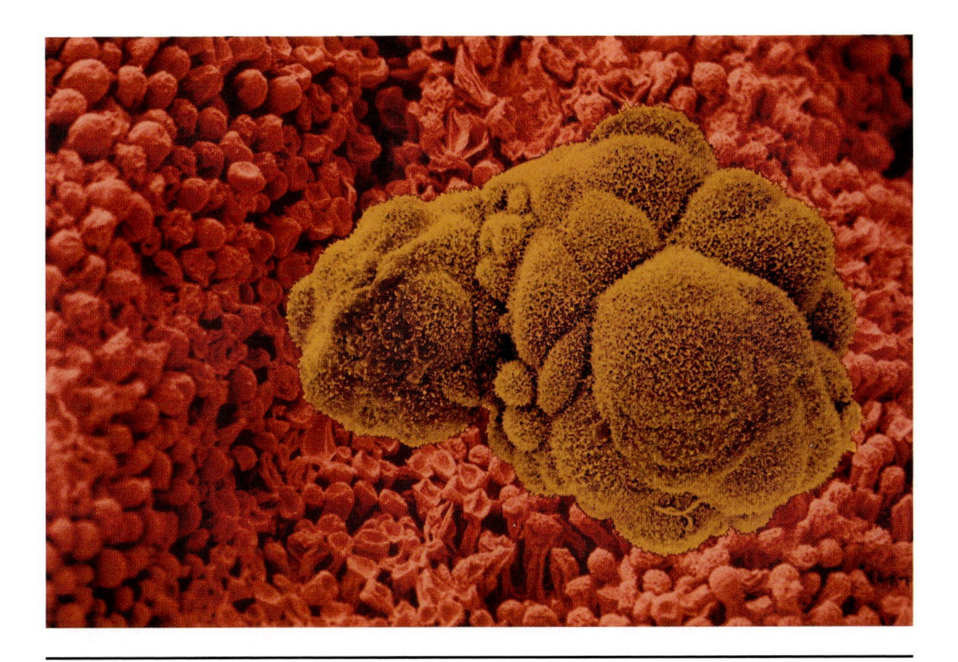

Developing blastocyst. © Dr. Y. Nikas/Phototake.

Dividing myoblast. © Dennis Kunkel/Phototake.

Muscle pain after exercise. © Hulton/Archive.

Marathon runner. © Skjold Photography.

Muscle hypertrophy from weight training. © Hulton/Archive.

A child with muscular dystrophy showing contractures. © Skjold Photography.

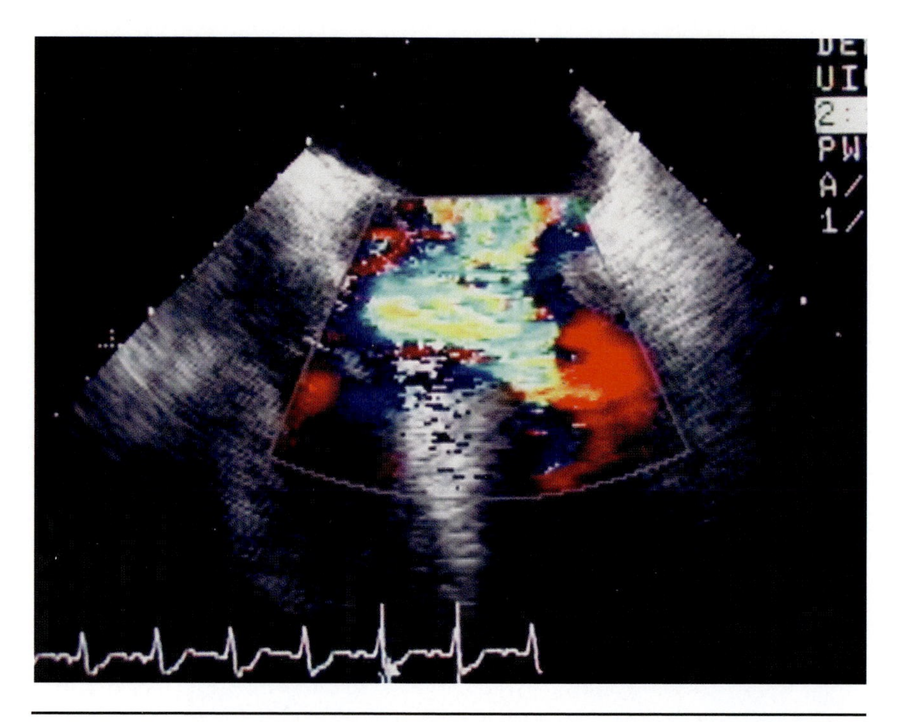

Echocardiogram showing hypertrophic cardiomyopathy. © Custom Medical Stock Photo.

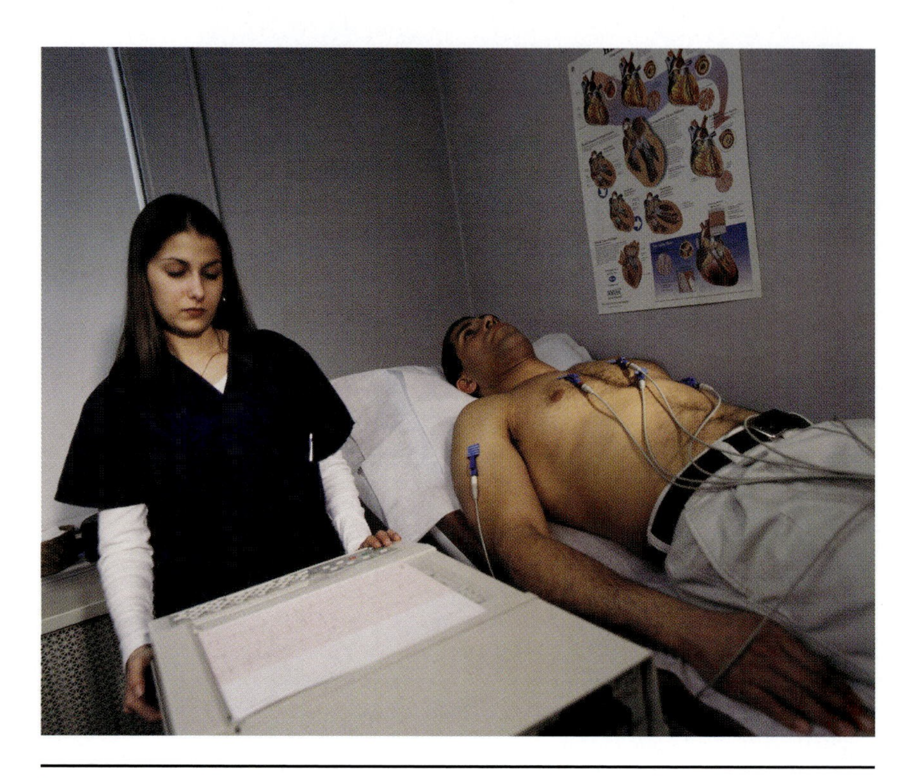

Electrocardiogram being performed on a patient by a technician. © Yoav Levy/Phototake.

8

Muscular Dystrophy

Muscular dystrophy is the second most common lethal genetic disease. The term "muscular dystrophy" actually refers to a collection of inherited diseases that all cause the muscles to slowly waste away. The diseases are lethal when the muscles that control breathing or the heartbeat can no longer function.

Each of the different muscular dystrophies is caused by mutations in genes that make muscle-related proteins. The most common form of the disease, called **Duchenne muscular dystrophy**, is caused by a mutation on the X chromosome. Because of how genes are inherited, males are much more likely to inherit Duchenne muscular dystrophy than girls. One in 3,300 male babies are born with Duchenne muscular dystrophy. The second most common form of the disease, called **Becker muscular dystrophy**, is closely related to the Duchenne form of the disease but tends to be less severe. Other much rarer forms of muscular dystrophy exist that are caused by mutations in other genes. These diseases are discussed in Chapter 9.

Despite a long history of research into muscular dystrophy, researchers have still not found a cure for the disease. Children who are diagnosed early can have physical therapy to help slow the muscle wasting or may be given drugs to help maintain some muscle strength. Neither of these approaches constitutes an actual cure for the disease—they slow the muscle wasting but don't prevent the disease from being lethal.

HISTORY OF MUSCULAR DYSTROPHY RESEARCH

Isolated cases of muscular dystrophy had been described since the 1500s, but it was Edward Meryon (1809–1880) who first studied the disease in detail. He based his research on a family with three sisters who all had male children with the disease, which they had gotten through three sisters. Meryon took muscle samples from the boys and found that their muscle fibers were completely destroyed and replaced with fat droplets and granular material. He concluded, correctly, that the disease affected the muscles themselves rather than the nerves.

The most famous scientist who worked on muscular dystrophy was Guillaume Benjamin Amand Duchenne (1806–1875). Duchenne muscular dystrophy, the most common form of the disease, was named after him. Duchenne's attention was drawn to the disease in 1856 when he saw his first case then went on to write about as many as forty different cases of the disease.

Duchenne's original description of the disease still holds true. He described the disease as a progressive weakening starting in the lower limbs then moving into the upper limbs. Despite this decreasing strength, some muscles—most notably the calves—grew larger as the disease progressed. He also noted that the disease usually started during adolescence and was more common in boys than in girls.

One major advance in Duchenne's work was his use of a device he called a needle harpoon—essentially a primitive biopsy needle—to take muscle samples from people with muscular dystrophy. Until he invented the device, researchers such as Meryon could only look at muscle samples in people who had died of the disease. By examining the muscle in people with various stages of the disease, Duchenne could learn how the disease progressed. His main finding was that in muscles of people with muscular dystrophy, the tissues surrounding the muscles slowly replaced the muscle fibers as they degenerated.

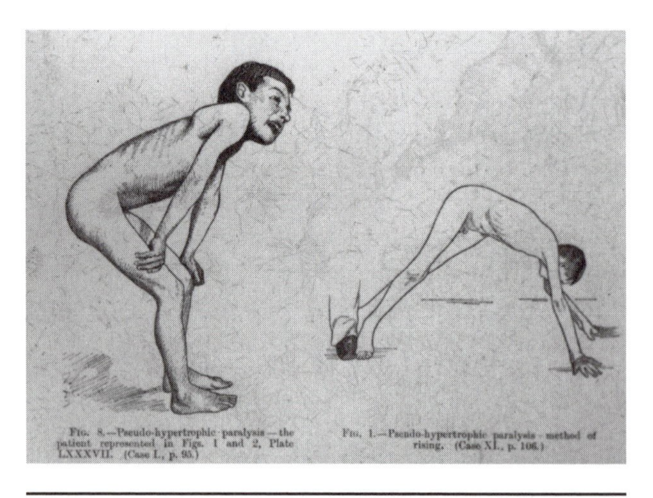

Child using Gower's maneuver to stand up. © National Library of Medicine.

William Gowers (1845–1915) is best known for describing the clinical signs of various diseases of the nerves and muscles. Among these is the **Gowers' maneuver**, which people with muscular dystrophy often use when getting up from a chair or from the floor. In the Gowers' maneuver a person with muscular dystrophy puts his or her hands on the floor with the weight on the toes, then slowly walks the hands toward the feet. To stand up, the person puts one hand on the knee with the other on the floor to push upright. This labor-intensive way of standing up compensates for extreme weakness in the legs and hips. Gowers later realized that this technique is not symptomatic of muscular dystrophy. Rather, it is used by any person with similar weakness.

In addition to studying signs of the disease, Gowers noticed that the disease was much more common in boys than in girls. He also observed that a woman could have sons with muscular dystrophy from different husbands, whereas husbands only had sons with the disease through one wife. This pattern of inheriting a disease through the mother was already known at the time through studies of a blood disease called hemophilia. Although nobody understood how that form of inheritance worked, it had been well described and was a known phenomenon. Figure 8.1 shows an example of how Duchenne Muscular Dystrophy is inherited in families.

Wilhelm Heinrich Erb (1840–1921) was the first to coin the term "progressive muscular dystrophy" to describe the disease that Duchenne and others had studied. In reviewing the literature, Erb noticed variability in the disease. It could begin at a range of ages and was much more severe in some people than in others. He was the first to suggest that muscular dystrophy was actually made up of several different diseases that all have similar symptoms. Since Erb first made this observation, researchers have learned the basis for several different forms of muscular dystrophy. All forms are inherited and cause more or less identical symptoms, but they progress at different rates depending on the **gene** that has a **mutation** in each form of the disease.

GENETICS OF MUSCULAR DYSTROPHY

The Dystrophin Gene

Muscular dystrophy is an inherited disease caused by a mutation in one of several genes. The gene that is mutated in Duchenne muscular dystrophy is located on the X chromosome. This gene makes a protein called **dystrophin**.

Genes on the X chromosome are inherited differently than genes on other chromosomes. All people have twenty-three pairs of chromosomes for a total of forty-six chromosomes in each cell. Two of these chromosomes are called sex chromosomes. Females have two X chromosomes and males have an X and a Y chromosome. When the sperm and eggs are formed, each sperm or egg gets only one of each chromosome. So, an egg will get one of the two versions of chromosome 1, one of the two versions of chromosome 2, one of the two versions of chromosome 3, and so on. Each egg also gets one of the two X chromosomes. In men, the sperm get either an X or a Y chromosome.

If a sperm containing an X chromosome fertilizes the egg, then the result-

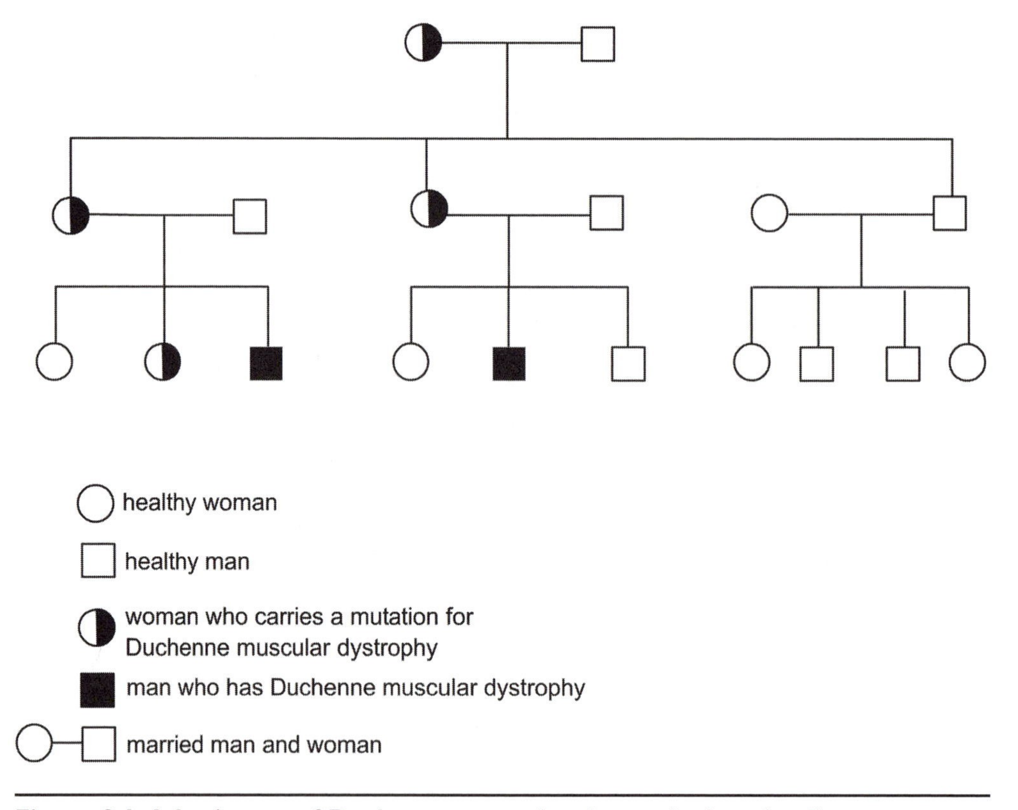

Figure 8.1. Inheritance of Duchenne muscular dystrophy in a family.

ing baby has an X chromosome from both parents and is a girl. If the sperm that fertilizes the egg has a Y chromosome, then the baby has an X chromosome from the mother and a Y chromosome from the father and is a boy.

The Y chromosome is very small and contains just a few genes that are necessary for the child to develop as a male. The X chromosome, on the other hand, has many important genes. This difference in the X and Y chromosome is why boys are more prone to several diseases including hemophilia and Duchenne muscular dystrophy.

If the mother has a mutation in the dystrophin gene on one of her two X chromosomes, she is called a **carrier** for muscular dystrophy. She doesn't develop the disease because she also has a normal form of the gene and can therefore make enough dystrophin protein for her muscles to function normally. However, she only has one gene making the protein and may therefore make less dystrophin than normal people. In some rare cases, women who were carriers will show some mild symptoms of muscular dystrophy as they get older.

If a mother who carries a mutation in the dystrophin gene forms an egg with the normal X chromosome, then the resulting baby will be normal. But if the mother forms an egg containing the mutated X chromosome, the baby will also carry that mutation. If the sperm that fertilizes the egg also has an X chromosome, then the resulting child will be a girl and will carry the mutation for Duchenne muscular dystrophy just like her mother.

However, if the egg containing the mutated X chromosome is fertilized by a sperm containing a Y chromosome, that baby will be a boy and will develop Duchenne muscular dystrophy. The father's Y chromosome had important genes that helped the child develop as a male, but the Y chromosome does not contain a normal copy of the dystrophin gene. The only dystrophin protein that child makes comes from the mother's mutated copy of the gene. This inheritance pattern is called **sex-linked inheritance** or, more specifically, "X-linked inheritance." Figure 8.2 shows how mutations in the dystrophin gene are inherited.

Another form of muscular dystrophy, called Becker muscular dystrophy,

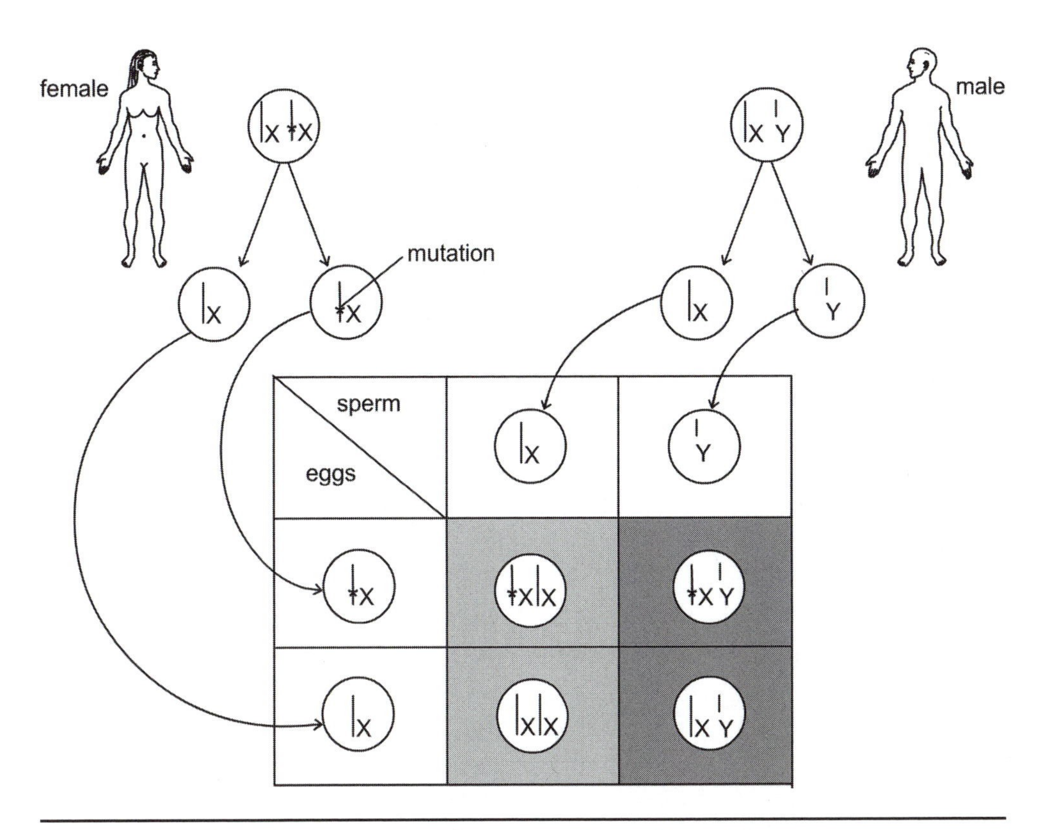

Figure 8.2. X-linked inheritance from an unaffected mother.

is also caused by a mutation in the dystrophin gene and is therefore also inherited in a sex-linked manner. This mutation happens at a different location in the gene and creates a dystrophin protein that is altered in a different way. Children with Becker muscular dystrophy tend to develop the disease later than children with Duchenne muscular dystrophy.

Although Duchenne and Becker muscular dystrophies account for almost all children with the disease, there are other genes that can be mutated to cause muscular dystrophy. These genes are contained on the nonsex chromosomes, also called **autosomes**. In order for a child to develop these forms of muscular dystrophy, both parents have to carry one copy of the mutation. If an egg that contains the mutated chromosome is fertilized by a sperm that contains the mutated chromosome, then the baby will have two mutated genes. That baby won't make any normal protein and will go on to develop muscular dystrophy. Because these forms of the disease aren't on the X or Y chromosomes, they are equally common in boys and girls.

The Dystrophin Protein

The most commonly muted gene in muscular dystrophy is a gene that makes the dystrophin protein. Dystrophin dots the outside of the sarcolemma—the membrane that surrounds a muscle fiber. It binds to other proteins to form a lattice that seems to support the fiber and protect it from injury.

Although researchers still are not sure exactly what role dystrophin plays, it appears to help keep the sarcolemma intact. Some symptoms of muscular dystrophy seem to be caused by leaky cell membranes. Enzymes from within the cell leak out into the blood, and chemicals such as calcium that are supposed to be kept out of the muscle fiber until it receives a signal to contract leak in. The way a signal sweeps down the muscle fiber from the junction between the nerve and muscle also relies on sodium and potassium being kept on opposite sides of the membrane. In people with muscular dystrophy, the signal fades out as it moves down the membrane, probably because the leaky membrane lets chemicals through rather than keeping them separate.

MOLECULAR CHANGES IN MUSCULAR DYSTROPHY

One question for researchers has been what exactly goes wrong in muscular dystrophy. Doctors know how to identify the disease, what mutations cause the disease, and what the dystrophin protein does in a muscle fiber, but they are still not sure exactly what causes the muscles to become weaker. Nor do they know why the disease affects some muscles more than others, or some portions of muscles before other portions.

One thing researchers do know is that the lack of dystrophin does not prevent the muscle from contracting normally. As stated in Chapter 6, researchers can create a skinned fiber by taking the membrane off of a muscle fiber to see how that fiber contracts under different conditions. When they create a skinned fiber that lacks dystrophin, that fiber contracts normally and the sarcoplasmic reticulum takes up calcium at normal rates.

Although the muscle is capable of contracting, some experiments suggest that the muscles don't mature normally. Muscles from people with muscular dystrophy still make some fetal forms of muscle proteins. These proteins usually stop being made late in fetal development, when the muscle switches to making mature forms of the protein. These fetal proteins include some that are involved in contraction, such as fetal myosin, as well as some enzymes that help the muscle make ATP. Although the failure to mature fully would make the muscle abnormal, that still does not explain why the muscles waste away in muscular dystrophies.

One possible hint at what lies behind muscular dystrophy comes from a study in muscle cells in a lab dish. When these cells, taken from people with Duchenne muscular dystrophy, should begin to produce the dystrophin protein, they accumulate high levels of calcium. This calcium is not present in muscle cells that make dystrophin at the normal time. It is still hard to interpret how and why calcium increases in the cell when dystrophin is absent, but it does point to a possible reason why muscles degenerate in muscular dystrophy.

SYMPTOMS OF MUSCULAR DYSTROPHY

Muscle Weakness in Babies

One of the earliest signs of muscular dystrophy is that a baby seems weak at birth, also known as floppy baby syndrome. This syndrome is also a sign of many other muscle and nerve diseases including cerebral palsy or mental handicap.

Other early signs of muscular dystrophy include a diagnosis of "failure to thrive." This somewhat vague diagnosis includes factors such as the baby not eating normally, not gaining weight, and not being as active as other babies. Children with muscular dystrophy also learn to talk or crawl at a later age than normal. In one study, doctors found that children with muscular dystrophy took about six months longer to learn to walk than other children. All of these signs could indicate a range of diseases, but together are symptomatic of muscular dystrophy.

Muscle Weakness in Young Children

Most children are not diagnosed with muscular dystrophy until they are around age 5 or 6. By this time parents have usually noticed that their child isn't able to run like other children, walks on his or her toes, sometimes throws out the leg when walking, and has a hard time climbing stairs. Children with muscular dystrophy are also more prone to falling than other children.

The muscles start getting weak in a consistent pattern. The weakness starts in the lower limbs, usually in the muscles that are closer to the center of the body such as the muscles in the thigh and buttocks. The hamstrings in the back of the thigh are much less affected by the disease. When the weakness spreads to the upper body, the muscles that are most affected are the triceps on the back of the arm rather than the biceps, and muscles that extend the wrist rather than the muscles that flex the wrist.

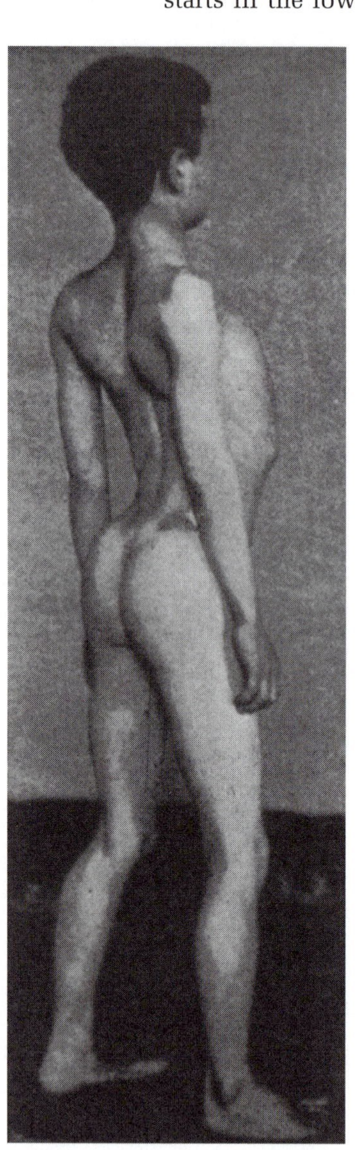

Typical posture in a child with Duchenne muscular dystrophy. © National Library of Medicine.

Even within a muscle, the disease is more apparent at one end of the muscle than the other. For example, in the pectoral muscles that cross the chest, the end that attaches to the center of the ribcage becomes weak before the end of the muscle that attaches to the shoulder. As the disease gets worse, these differences in which muscles are weakest or which portions of muscles are weakest go away, and all the muscles are equally damaged. At this late stage of the disease, the face muscles may also get weak and start to droop. Even as other muscles in the body become less able to move, muscles that are involved in chewing and swallowing are never affected.

Physical Effects of Weakened Muscles

Children with muscular dystrophy have a very characteristic way of walking and moving because of their weaker muscles. Because of which muscles become weak first, the children compensate by moving their legs differently as they walk. For example, to compensate for the weak buttocks muscles, the child tilts the entire pelvis toward the lifted leg with each step. This motion creates a waddling walk, which is also seen in other muscle or nerve-wasting diseases.

Children with muscular dystrophy also tend to stand and walk with their pelvis tilted down in the front, creating an arched back (see photo). To compensate for this, the lower back takes on a characteristic shape, with a deep dip just above the buttocks. All of these changes in the back and in how the legs move tend to throw the child off balance. Walking on the toes seems to help retain the balance and also uses the stronger muscles in the feet.

When a person stands up from the ground, straightening the legs uses muscles in the buttocks and hip, neither of which are strong in children with muscular dystrophy. To make up for this, they use the characteristic Gowers' maneuver to stand up. One way doctors test this is by asking a child to stand up from the floor with his or her arms crossed. Normal children have no trouble doing this, whereas even children with very early muscular dystrophy find it to be impossible.

There are so many affected muscles that it has been hard for researchers to pick apart the individual reasons for all aspects of the unusual walking gait of a child with muscular dystrophy. Regardless of the reasons behind each aspect of how the child walks, stands, or moves, it does add up to a very characteristic pattern that can be used to diagnose muscular dystrophy.

Enlarged Muscles

Although muscular dystrophy is a disease in which the muscles slowly waste away, one of the early signs of the disease is enlarged muscles. In this case, larger muscles are not stronger muscles. Instead, the muscles are larger because the tissues surrounding the muscle expand and fat builds up within the muscle. This extra tissue often makes the muscles feel firm and may be described as "woody."

The most commonly enlarged muscle in muscular dystrophy is the calf, though in some cases muscles in the back and shoulders may also be affected. Although almost all people with muscular dystrophy do have enlarged calves, if not other muscles, this sign is not unique to the disease and can't be used alone as a diagnosis.

PROGRESSION OF DUCHENNE MUSCULAR DYSTROPHY

Early Progression

Muscular dystrophy becomes progressively worse as a child gets older, but seems to get worse in spurts. For a while the muscles will remain about the same, then suddenly they will become much weaker and new muscles will start to be affected. This fluctuation has made it hard to evaluate therapies for the disease, because a therapy may coincide with a period of time when the disease is naturally not getting worse. Researchers generally have to watch the child for many years in order to tell whether the therapy was effective.

As the muscles become weaker, the lower back becomes more and more swayed and the waddling walk gets more severe. Eventually, the muscles get so weak that it is not possible to compensate by waddling and the child has to use a wheelchair. At first, children will only use the wheelchair for longer trips, but inevitably the disease becomes severe enough that the child can't get around without it. Most children with muscular dystrophy are using a wheelchair by about age 12. In general, the earlier the child is confined to a wheelchair, the earlier in life that child dies.

Not only do children with the disease get weaker at different ages, but they also tend to look different. Some children keep their muscle bulk and still have fat under the skin, even though the muscles themselves are extremely weak. In other children who are at a late stage of the disease, the muscles waste away almost completely and the children have very little fat under the skin. Although kids vary quite a bit in how their muscles appear superficially, siblings with the disease tend to follow a very similar pattern.

Late Progression

As the disease continues to get worse, the muscles stiffen and hold the limbs in an unusual shape, usually holding the elbows and knees straight and the feet pointed. This stiffening, called **contractures**, also limits how far the child can move the arms and shoulders.

The most serious deformity comes when the muscles around the chest contract and pull on the ribs, spine, and shoulders. These contractions can be so intense that they deform the bones and squeeze the heart and lungs. Eventually, the lung size gets smaller and the child must take shallower breaths. This effect is made worse by the fact that the muscles in the ribs that are used in breathing get weaker, so the child has a harder time getting a full breath of air. When the disease gets to this stage, the child often has less oxygen in the blood than is normal, so not enough oxygen gets to all cells in the body. Late in the disease, the heart muscle gets weak and can't pump blood normally.

Most children die from breathing problems. Either their lungs are no longer able to get enough air or the weakened lungs contract pneumonia, which is the eventual cause of death. Rarely, a child may die from having an irregular heartbeat. Overall, most children die from muscular dystrophy by age 20, and they may die as early as age 10. Despite the past 100 years of intense study and several attempted treatments, this average age of death is no different from what Gowers saw back in the late 1800s.

PROGRESSION OF BECKER MUSCULAR DYSTROPHY

Becker muscular dystrophy is less common than Duchenne muscular dystrophy and tends to be less severe, although it is caused by a mutation in the same gene. The difference between the two diseases appears to be what type of mutation occurs. In Duchenne muscular dystrophy, the mutation is so severe that no functional dystrophin protein is made. In the Becker form of the disease, the mutation is such that some dystrophin protein is made, although it is altered and not as functional as the normal protein. Because people with the Becker muscular dystrophy have some functional dystrophin, they are not as severely affected by the disease.

People with the Becker form of the disease have wasting in the same muscles as in Duchenne's and the muscles become weak in the same order, with the lower limbs affected first, followed by the upper limbs. Both forms of the disease share the symptom of enlarged calves.

OTHER X-LINKED MUSCULAR DYSTROPHIES

Duchenne and Becker muscular dystrophies are by far the two most common sex-linked muscular dystrophies. However, researchers over the past fifty years have studied isolated families with sex-linked muscular dystrophies that progress differently from either the Becker or Duchenne form of the disease. These muscular dystrophies follow a similar progression as Duchenne muscular dystrophy but may affect different muscle groups, progress slower, or have more problems with the heart. These forms of the disease are discussed in Chapter 9.

DIAGNOSING MUSCULAR DYSTROPHY

It used to be that if doctors noticed early signs of muscular dystrophy, they would simply have to watch the child develop and look for later signs to confirm a diagnosis. Now doctors have several tools at their disposal to determine which form of muscular disease a child has. It is especially important to be certain that a child has muscular dystrophy versus other diseases because some diseases that may appear like muscular dystrophy, such as polymyositis, are treatable if diagnosed correctly. Figure 8.3 shows the steps a doctor goes through in diagnosing muscular dystrophy.

Creatine Kinase

One of the easiest ways of testing for muscular dystrophy is to sample the level of creatine kinase in a child's blood. Remember that creatine kinase leaks out into the blood from muscles damaged by difficult exercise (Chapter 7). In 1959, researchers learned that muscles being damaged by muscular dystrophy also release creatine kinase. This could be because the sarcolemma is leaky in people without sufficient dystrophin and allows muscle proteins to escape.

When doctors know to look at creatine kinase in very young babies with Duchenne muscular dystrophy—possibly because the child's sibling also had the disease—they find that babies have levels that are as much as 100

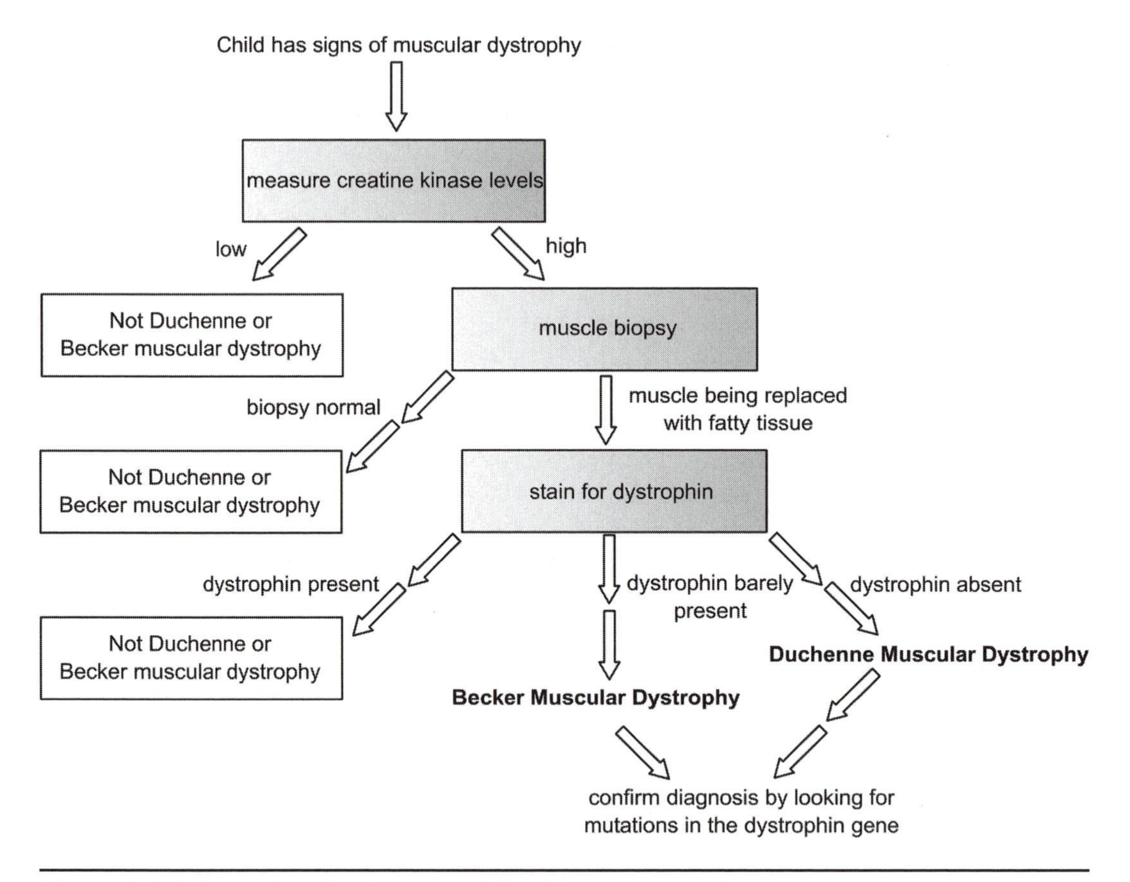

Figure 8.3. Diagnosing Duchenne and Becker muscular dystrophy.

times higher than normal babies. In fact, one study found that fetuses with muscular dystrophy already have high creatine kinase levels in the last third of the pregnancy.

These studies show that although the disease doesn't become apparent until much later, the muscle wasting starts at an early age. The creatine kinase levels are extremely elevated in children with muscular dystrophy until about age 6. At that time the levels start going down to normal, and they are near normal after the child is confined to a wheelchair. Although researchers aren't exactly sure why this change takes place, it could be because the muscles have wasted so much that there is little creatine kinase left in the muscle to leak out. It could also be that the creatine kinase seeps out of the muscle after the muscle has been used walking or playing. When the child is confined to a wheelchair, the muscles are usually too weak to use at all, much less to damage while playing. In children with the Becker form of the disease, the creatine kinase levels don't start dropping until about age 15, which coincides with the slower progress of that disease.

Electrical Recordings

When muscles contract they create electricity. This electricity varies depending on how many muscle fibers were involved in the contraction, how strong the contraction was, how long it lasted, and where on the muscle the researcher recorded the electricity. Although it isn't possible to diagnose muscular dystrophy based on electrical readings, it can give doctors one more piece of evidence to help confirm the diagnosis. Electrical recordings have also helped researchers start to piece together the molecular defects in muscular dystrophy.

The electrical reading is usually taken using a fine needle inserted into the muscle. Given how many children feel about needles, this test can be difficult for all but experienced doctors to give. However, when it is done right, signals from the needle displayed on a computer monitor can show individual muscle fibers contracting.

In children with muscular dystrophy, the single contractions don't last as long as in healthy children and the signal to contract isn't as strong. In children with the disease and in mice that lack dystrophin protein, the signal to contract seems to dissipate as it moves down the muscle. A signal that is strong at the junction between the nerve and muscle in the middle of the fiber may not reach the far ends of the fiber.

One reason why the signal for each contraction isn't as strong in muscular dystrophy is because each motor unit is smaller. When a single nerve stimulates the muscle to contract, fewer fibers respond. Eventually, as the muscle fills with fatty deposits, some nerves don't go to any functioning muscle fibers and researchers record no activity from that part of the muscle.

As with so many other signs of muscular dystrophy, electrical recordings of the muscles as they contract are not characteristic of muscular dystrophy and can't be used to distinguish between different forms of dystrophies. However, if the recordings are done carefully they can scale down the number of possible diseases that may be responsible for the muscle weakness.

Muscle Biopsy

It turns out that although muscular dystrophy is a wasting disease that doesn't become readily apparent until later in childhood, the muscles look abnormal under a microscope even at a very early age. As the wasting progresses, the muscles look less and less normal, making a biopsy a good way to assess how far the disease has progressed.

EARLY STAGES

In the earliest stages of muscular dystrophy, the muscle fibers are much more varied in size than normal muscles. Some fibers also become very dark in response to a common muscle stain called eosin. These fibers that stain darkest in response to eosin also have very high levels of calcium. Chapter 7 stated that high levels of calcium can damage the muscle by preventing the mitochondria from making ATP. Researchers think that the fibers with high levels of calcium are at the early stages of breaking down.

This eosin staining is one of the few ways to distinguish between the different forms of muscular dystrophy. In Duchenne muscular dystrophy, muscle biopsies contain many of these dark-staining, high-calcium fibers. Becker muscular dystrophy also shows high levels of eosin-staining fibers, whereas they are rare in other forms of the disease.

Early in muscular dystrophy, some of the muscle fibers appear to be undergoing repair. These fibers have a smaller cross-section and have large, pale nuclei. Over time, the biopsies contain fewer and fewer regenerating fibers and more fibers that are disintegrating.

LATER STAGES

In later stages of the disease, muscle fibers swell and lose their characteristic stripes, which indicates that the sarcomeres have broken down. The nuclei within the fibers disintegrate, and the fibers look waxy from the muscle proteins that clump together. Eventually, immune cells from the blood enter the fiber and clean up the dying tissue.

Where muscle fibers have broken down, fatty tissue takes over the space. In biopsies of people with later stages of muscular dystrophy, the muscle fibers themselves appear like islands in the fatty tissue. This progression in how muscles look under the microscope is characteristic of muscular dystrophy and not of other diseases. It is one of the best ways to confirm a diagnosis.

Dystrophin Staining

Most forms of muscular dystrophy involve the gene that makes dystrophin. With that in mind, one effective way of diagnosing the disease is to look for dystrophin in a muscle biopsy. Chapter 6 discussed that researchers can make antibodies that specifically recognize a given protein. They can also make those antibodies carry a fluorescent marker that is visible under a microscope.

When researchers stain a muscle biopsy sample with an antibody against dystrophin, they see very different patterns in normal muscle versus muscle from a person with muscular dystrophy. In a normal muscle biopsy, the antibody forms fluorescent rings around the normal fibers. Biopsies from people with Duchenne muscular dystrophy show almost no antibody staining whatsoever. These fibers have no dystrophin on their membrane surface. In Becker muscular dystrophy, the fibers have some dystrophin on the surface, which shows up as faint antibody staining surrounding each fiber.

Mutations in the Dystrophin Gene

The most conclusive way to diagnose a case of Duchenne or Becker muscular dystrophy is to look for specific mutations within the gene. If a doctor does not find a mutation, that child still may have the disease but with a mutation in a different gene. This test will diagnose the large majority of people with the two most common forms of the disease.

To find mutations in dystrophin, doctors draw blood and remove DNA from the white blood cells (red blood cells do not have DNA). They then use a technique called Polymerase chain reaction (PCR) to make thousands of copies of the dystrophin gene. Researchers can then use another technique called gel electrophoresis to examine the size of the dystrophin gene. In gel electrophoresis, researchers look at how quickly that gene moves through a clear jello-like slab in response to electricity. Larger molecules move slowly through the gel, while smaller molecules move more quickly.

Most mutations to the dystrophin gene involve deletions of large portions of the gene. On a gel, these mutated genes move faster than normal forms of the gene because they are much smaller. The deletion in the gene also deletes part of the protein itself, which completely prevents the protein from functioning normally.

An even more precise way to diagnose the disease is to make many copies of sections along the gene. If the gene is divided into ten sections and the deletion is in the second section, then all sections will look just like the normal gene on a gel except for section 2. This lets researchers pinpoint exactly where that mutation is located. Later, if parents want to see if a fetus has the disease, doctors will know that the mother carries a deletion in section 2 of the gene, and will only need to look for mutations in this particular section.

Using PCR to make many copies of the gene only detects mutations if the gene changes size—such as if there is a deletion or a duplication of a region of the gene. In about 30 percent of cases, the mutation only causes a switch at a single location in the gene, which won't show up on a gel. This switch will leave the gene the same size, although the protein that is made may be completely different and unable to function. Researchers are able to detect these mutations by directly sequencing the gene, although the technique takes quite a bit longer and is more involved than using a gel to discover deletions.

MANAGING MUSCULAR DYSTROPHY

Despite everything that is known about muscular dystrophy, researchers still do not have a cure for the disease. In the past, when a child was diagnosed with muscular dystrophy there was little a doctor could do to help the child. Now, there are several ways to make a child more comfortable and to slow the disease progression, although the disease is still always lethal.

Active Exercise

While a child is still able to move around normally, there has been debate about whether the child should participate in active exercise. One argument has been that exercise may strengthen the muscles, delaying their wasting. In one study, boys who had recently been diagnosed with muscular dystrophy lifted weights regularly for a year. Despite some early gains in strength, the boys still became progressively weaker. The exercise did not seem to make the disease any worse or progress any faster. Some doctors have suggested that because exercise tends to help people keep their spirits up, some form of regular exercise may be good for the children emotionally if not physically. Swimming is often recommended because it puts very little stress on the muscles.

Passive Exercise

In addition to the muscles becoming weaker, children with muscular dystrophy often have muscles contracted into awkward and uncomfortable positions. Many studies have shown that stretching or having another person move the joint can significantly delay these contractures. Doctors recommend physically moving the limbs around in the evening after a bath when the muscles are most likely to be relaxed. Once a child develops contractures, however, stretching can actually further damage the muscle and should be stopped.

Back Supports

Once a child loses the ability to walk and is confined to a wheelchair, the spine often curves in a form called scoliosis. With scoliosis, sitting is difficult if not impossible and the curved back squeezes the lungs, making it hard to get a full breath of air. This can aggravate problems with the chest muscles that already make breathing difficult for a child.

One way to prevent scoliosis is to fit a child with a back brace to hold the spine straight and the back upright. This can delay any curvature of the spine but does not prevent it. Once scoliosis has developed, the only way to reverse it is through surgery.

In children with Duchenne muscular dystrophy, surgery to hold the back straight can improve the quality of life somewhat, but does not necessarily slow the eventual progress of the disease. However, in people with milder forms of Becker muscular dystrophy, who are often still mobile into their 20s or 30s, spinal surgery to hold the back straight may make a big difference in the person's ability to maintain their independence or keep a job.

Keeping a Child Walking

When a child becomes confined to a wheelchair, scoliosis and contractures usually soon follow. If a child walks as long as possible, it seems to delay these two side effects of the disease. In the 1960s a group of researchers developed full-leg braces that support the legs and hips and allow a child to support him or herself even when the muscles are too weak on their own. These can be adjusted as the child grows. In addition to the leg braces, most children also need to have operations on the tendons to prevent them from tightening and deforming the leg. Once a child is fitted with the braces, they can keep the child out of a wheelchair for an average of about two years.

Keeping a child walking has many benefits in addition to delaying contractures and scoliosis. The children seem happier when they can be somewhat independent, and they are much easier for a parent to take care of. Despite these benefits, delaying the time when a child must eventually use a wheelchair does not seem to delay the average age of death.

GENETIC TESTING

Parents who know they have muscular dystrophy in the family may want to have genetic testing to find out if they run the risk of having children with the disease. The first step in genetic testing is to find the mutation in a family member with the disease. Each family with muscular dystrophy has a slightly different mutation. Usually that mutation will be within the dystrophin gene. However, in some cases the mutation is outside the gene itself but still prevents the gene from making dystrophin protein. The family may also have a form of muscular dystrophy that does not involve the dystrophin gene. A test that looks for a mutation in the dystrophin gene in this family would come back negative, even if a woman does carry the disease mutation.

If the doctor does not find a mutation in the family member, then that family has a mutation that cannot be located by current technology and genetic testing is not appropriate for the couple. This situation comes up very rarely. If the doctor does find a mutation in the family member, then the woman can be tested to see if she also carries that mutation. If she does, then her sons have a 50 percent chance of inheriting her mutated X chromosome and having muscular dystrophy.

Genetic testing can only give a couple information about whether they run the risk of having children with the disease. Each couple has to decide what to do with that information. Some couples may decide to get pregnant, then have the fetus tested for the mutation. If the fetus is a boy and has the mutation, the couple may decide not to have the child. Other couples may decide to adopt rather than run the risk of having children with a disease that is currently incurable.

TREATING MUSCULAR DYSTROPHY

Drugs

Over the past fifty years, many different drugs have been tested to treat muscular dystrophy. These include drugs to protect muscle cells from calcium buildup, minerals to stabilize the muscle fiber membrane, and steroids to maintain muscle mass. Of these, the only drugs that have proven to be even somewhat effective are steroids. Researchers first thought of using these drugs after a child with Duchenne muscular dystrophy had a kidney transplant and had to go on an immune-suppressing steroid called prednisone to avoid rejecting the new kidney. This child's disease progressed more slowly than is normal.

In trials with many more children, prednisone seems to slow down the muscle wasting. However, the drug has unpleasant side effects that are not necessarily an improvement over weakening muscles. Steroids can make a child susceptible to cancer, slow growth, and infections because the immune system is suppressed. Prednisone can also cause osteoporosis, headaches, and increase child's risk of developing diabetes.

Transplanting Muscle Cells

If muscular dystrophy causes muscle fibers to waste away, then perhaps new muscle cells can be injected to replace those that are lost. When Helen Blaw tested this idea in people with muscular dystrophy in 1992, they found some healthy, intact muscle fibers making dystrophin near the site of the injection. Since that time, the challenge has been getting those implanted cells to spread to muscles throughout the body and to prevent the immune system from rejecting the transplanted cells. Both of these hurdles have been difficult to overcome.

The most effective way of transplanting new muscle cells has been using satellite cells. Researchers take a muscle sample from a person with healthy muscles and isolate satellite cells. They can then put those cells in a lab dish where they divide to make many copies of themselves. Once enough new cells have formed, the researcher can inject those cells into muscles throughout the body. The satellite cells divide, fuse with damaged fibers, and help repair the muscle.

In some early trials in humans, transferring normal satellite cells into damaged muscles both prevented those muscles from becoming weaker and also seemed to make the muscles somewhat stronger. The muscles looked more normal under a microscope, and some of the fibers were surrounded by dystrophin. Although these results were promising, this therapy involves injecting satellite cells into every affected muscle in a child's body. This is both time consuming and requires more satellite cells than can easily be grown in a lab dish.

Gene Therapy

By far, the most common forms of muscular dystrophy are due to mutations in the dystrophin gene. Fixing that gene using gene therapy would seem like the ideal solution for treating the disease. The idea behind gene therapy is that a good copy of the gene can be inserted into the tissue, then this gene incorporates into the cell's own chromosomes. The gene then makes normal protein and the disease is cured.

Although the idea behind gene therapy is straightforward, researchers have encountered many hurdles in getting gene therapy to work for any disease, much less muscular dystrophy. The first problem is getting the gene into the cell's chromosomes. If a person were to simply inject DNA into a cell, the DNA would be broken down as a foreign molecule. To get around this, most gene therapy approaches use viruses, whose normal role is to insert their DNA into cellular chromosomes.

Viruses that are used in gene therapy have most of their own genes removed and the one gene of interest—in this case, dystrophin—inserted. The viruses only have enough genes to make proteins needed to insert the DNA into a chromosome. Researchers can inject this modified virus into the tissue and have the virus insert the therapeutic gene into a random location on the chromosome. In some cases, that gene will then make a normal protein.

There have been some limited successes using viruses to deliver genes in gene therapy. The main problem with the approach is that the gene inserts at a random location. Sometimes that random location is one where not much protein can be made from the gene. In a French gene therapy trial for an unrelated disease, the gene inserted in such a way that it disrupted another gene, causing a mutation that led to leukemia.

Some newer gene therapy approaches are currently being tested for muscular dystrophy. These involve a different way of having the gene insert into a chromosome without the aid of a virus. With this approach, the gene enters the chromosome at a known site that is not near other genes that could be mutated. Even if this approach turns out to be safe, the question remains of how to get the good form of the dystrophin gene into all muscle cells. An injection into the arm or leg will only get the gene into nearby tissues, whereas muscular dystrophy affects muscles throughout the body, including the heart muscle.

One approach to getting the dystrophin gene throughout muscles of the body has been to inject the gene in an appropriate form, virus or otherwise, into the blood supply. This will take the gene to muscles and other tissues throughout the body. The dystrophin gene will integrate into chromosomes in a wide range of tissues but will only make protein in muscle or other tissues where the protein is normally made.

Even if injecting into the blood supply distributes the dystrophin gene properly, problems remain. Doctors can only try gene therapy after a child has been diagnosed with the disease. If a doctor and the parents are very observant, they may detect the disease when the child is very young and still has most of his or her muscle mass, but in some cases the child is already quite weak by the time muscular dystrophy is diagnosed.

A functioning copy of the dystrophin gene, even if it successfully makes dystrophin protein in all the muscles, will only preserve the muscles that are left. The gene on its own cannot restore lost muscle fibers. In addition, doctors do not know whether the dystrophin gene may be needed during development. It is possible that giving the dystrophin gene during childhood will miss the critical period when dystrophin was needed to make fully functioning muscles.

9

Other Diseases of the Skeletal Muscles

Muscles are an extremely complex piece of machinery, with intricate protein structures that cause a contraction, a network of surrounding proteins to support the muscle fiber, membranes that transmit electrical signals throughout the fiber, and mitochondria that carry out the critical task of making ATP. Any change to one of these components can cause a muscular disease. Given how complicated muscles are, it may be surprising that muscular diseases are one of the rarest classes of disease. Muscle is also unusual in its resistance to cancer. Of all the tissues in the body, muscle is one of the least common places to develop a tumor.

The most common muscle diseases are Duchenne and Becker muscular dystrophy, as described in Chapter 8. But many other types of muscular dystrophies and other muscle diseases also exist, each with its own pattern of muscle weakness. Although they are rare, these diseases can be completely disabling. When the muscles don't work properly, a person has a harder time getting around, eating, or in some cases even breathing.

Diseases of the skeletal muscles can be broken down into several categories: muscular dystrophy, **myotonic dystrophy**, **myopathy**, metabolic diseases, and neuromuscular disease. Many of the neuromuscular diseases are more a problem with the nerves than with the muscles and are addressed in the Nervous System volume in this series. This chapter deals specifically with diseases of the skeletal muscles, including the rarer forms of muscular dystrophy.

MUSCULAR DYSTROPHIES

Chapter 8 described the most common forms of muscular dystrophy, both of which are caused by mutations in the dystrophin gene. Although these two forms make up the majority of all cases of muscular dystrophy, there are other forms of this disease caused by different mutations. All of the muscular dystrophies tend to have unique patterns of muscle weakness, have different ages of onset, and vary in how severe the disease becomes. Although each disease is unique, the dystrophies are united in causing at least some of the skeletal muscles to slowly waste away.

All of the dystrophies have some features in common, which can make it difficult to diagnose the exact form of dystrophy a child has. An exact diagnosis cannot help a doctor treat the disease, because no treatment exists for any form of muscular dystrophy. However, a correct diagnosis can help the parents know how quickly to expect the disease to progress and what to expect as the child gets weaker. Knowing exactly what diseases run in the family can also help other family members get appropriate genetic counseling and possibly get genetic testing to learn their own risk—or their child's risk—of the disease.

The first step in diagnosing a muscular dystrophy is to look at the disease in other family members. Because the dystrophies are all inherited, a doctor can often work out what type of disease a relative has and assume the child has the same form of the disease. If that doesn't work, then the doctor has to first look at which muscle groups are affected and whether the child has enlarged calves, which is a hallmark of Duchenne and Becker muscular dystrophies as well as some other forms of the disease. On its own a physical examination can rule out some forms of muscular dystrophy, but many of the diseases still appear very similar, especially in the early stages. In the end, a doctor may have to do a muscle biopsy to figure out which type of dystrophy a child has.

Limb Girdle Muscular Dystrophy

The **limb girdle muscular dystrophies** cause the muscles of the stomach, shoulders, and back to become weak. Sometimes people who inherit this form of muscular dystrophy do not start showing signs of the disease until late in life and never have symptoms worse than muscle weakness. Other people start becoming weak as children, and the symptoms soon spread throughout the body. Despite this diversity, most people with the disease start showing symptoms by the time they are in their 20s.

Research to learn more about genes causing limb girdle muscular dystrophy has been slowed down by the fact that each gene mutation can cause a wide range of symptoms in different family members with the disease. It has been hard for researchers to pick out people with the same form of the disease in order to look for common mutations. In the past, this disease was sometimes misdiagnosed as Duchenne or Becker muscular dystrophy.

GENETICS

So far, researchers have found mutations in fourteen genes that can cause limb girdle dystrophies. Eight of these are **dominant mutations** and the other six are **recessive mutations**. In a dominant pattern of inheritance, a person only has to inherit one mutated copy of the gene in order to develop the disease. In a recessive mutation, a person must inherit two mutated copies of the gene before they develop the disease.

Although mutations in genes that cause limb girdle muscular dystrophy are rare in the general population, they are much more common in some inbred populations such as the Amish and Mennonite communities. Gene mutations are usually common in a certain population because of a phenomenon called the **founder effect**. Within the group of people who founded those communities, one or more had a mutation that causes limb girdle muscular dystrophy. That person had children and passed the gene on to the next generation. Because people in that population tend to marry amongst themselves, the mutant gene remained common in the population even though it is a rare mutation outside that community.

Of the genes that can be mutated in limb girdle muscular dystrophy, almost all of them interact in some way with dystrophin—the gene that is mutated in Duchenne and Becker muscular dystrophies. Dystophin is a long, fiber-like protein that attaches to the muscle fiber membrane through a network of supporting proteins. These proteins anchor the dystrophin in place or attach to proteins on the other side of the membrane. Together they make up a complex web of proteins on the inside and outside of the cell membrane that stabilize the membrane and relay signals. When one of these proteins is abnormal, the entire network cannot function.

Although each of the proteins is part of the same network, they all have unique roles and cause slightly different forms of muscular dystrophy. These differences mainly have to do with age of onset, which can range from early childhood to well into the 20s, and in the specific muscle groups that are affected. The different forms of the disease also vary in how quickly the muscles become weaker.

IDENTIFYING THE DISEASE

Limb girdle muscular dystrophy is most often confused with Duchenne or Becker muscular dystrophy because many of the same muscle groups are affected, and in all three forms of the disease a child usually has enlarged calf muscles (see photo). These forms of muscular dystrophy also don't usually involve the face muscles, which can distinguish them from other forms of the disease. To confirm a diagnosis of limb girdle muscular dystrophy, a doctor usually has to do a muscle biopsy then look at how antibodies to dys-

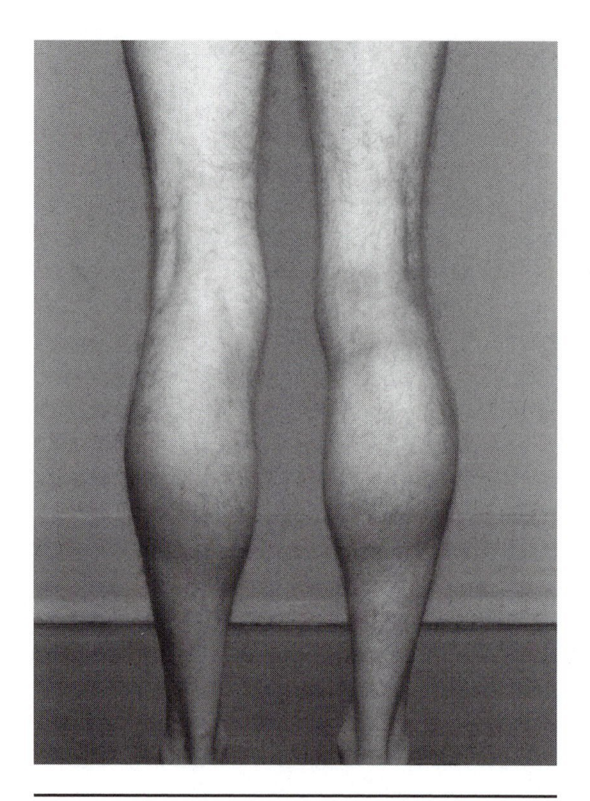

Calf hypertrophy in a child with limb girdle muscular dystrophy. © NMSB/Custom Medical Stock Photo.

trophin stain that sample. In Duchenne and Becker muscular dystrophies, the antibodies don't bind to any dystrophin so no fluorescent signal shows up under a microscope. Dystrophin is normal in limb girdle muscular dystrophy, so antibody staining for that protein looks like normal muscle.

Facioscapulohumeral Dystrophy

Facioscapulohumeral dystrophy (FSHD) affects the face (facio); scapula, or shoulder blades (scapulo); and humerus, or upper arms (humeral). It is the third most common form of muscular dystrophy after the Duchenne and Becker forms of the disease. Of all the muscular dystrophies, this form progresses the slowest and people with the disease usually live a normal lifespan. Only about 20 percent of people with the disease end up needing to use a wheelchair.

When people are diagnosed with facioscapulohumeral dystrophy, usually by 20 years of age, they sometimes remember that they have always had a hard time whistling or drinking through a straw. These mild symptoms can be an early sign of the disease. People with FSHD also have trouble doing push-ups or climbing rope due to weakness in the shoulder blades. By the time the disease is usually diagnosed, the shoulder muscles are so weak that the shoulder blades stick out like wings. People also usually trip frequently and have a hard time running.

FSHD usually follows a slow, steady progression. The muscles do continue to get worse, but at such a slow rate that people can often come up with ways to deal with the weakness and live a normal, productive life. Some people who have serious weakness in the calves may opt to use ankle braces that support the lower leg, but other people find that these make it even more difficult to walk. In people whose shoulder blades become extremely weak, doctors can perform a surgery that holds the shoulder blades in place. This can make it easier to lift the arms up but can also limit a person's range of motion. Some people opt to have only one shoulder blade operated on, so they end up with one arm that can lift objects without sacrificing their range of motion in the shoulders. Both the leg braces and the surgery can help for a short while, but they do not prevent the muscles from continuing to get weaker.

GENETICS

FSHD is caused by a dominant mutation that can affect both men and women. The mutation involved is highly unusual. In most diseases, mutations that cause the disease are located within a gene. This type of mutation causes the protein made by that gene to become altered or to not be made at all. Other common mutations are located near the gene in a DNA regulatory region that controls how much protein that gene makes and where that protein is made.

The mutation that causes FSHD is in a region of the chromosome where there are no nearby genes. Researchers do know about some instances where regulatory DNA regions can be a large distance from a gene. They think the region that is mutated in FSHD somehow controls distant genes. However, because that gene is not near the mutation, they don't know which gene it regulates and therefore don't know what protein is altered in the disease. Researchers are now using sophisticated genetic techniques to track down the gene that is affected by the FSHD mutation.

IDENTIFYING THE DISEASE

FSHD is one of the only muscular dystrophies that involves the face muscles, making this form easy to diagnose in most people. The calves usually grow weaker in this form of the disease rather than enlarging, which rules out Duchenne, Becker, or limb girdle muscular dystrophy. Because it is inherited in a dominant fashion, a person with symptoms usually has a relative with the disease, making it easier for the doctor to diagnose. Doctors rarely have to take a muscle biopsy to diagnose FSHD.

Emery-Dreifuss Muscular Dystrophy

Emery-Dreifuss muscular dystrophy usually starts out with weakness in the upper arms. People with the disease develop the unusual muscle contractions called contractures very early in the disease. These contractures usually start in the elbows and Achilles tendon and can also happen in the back of the neck. Although many forms of muscular dystrophy include contractures, this is the only form in which the contractures start while the muscles still have some strength.

Another hallmark of Emery-Dreifuss muscular dystrophy is **cardiomyopathy**, with an extremely high risk of **sudden death**. In this form of cardiomyopathy, the muscles mainly get weaker in the upper portion of the heart, called the **atrium**. Biopsy samples show that the muscle fibers of this region of the heart break down while fatty tissue takes over. People who are diagnosed with Emery-Dreifuss muscular dystrophy have such a high risk of sudden death that they often get a pacemaker to keep the heart beating steadily whether or not they already show signs of cardiomyopathy.

GENETICS

So far researchers have identified three different genes that can be mutated in people with Emery-Dreifuss muscular dystrophy. The most common gene to be mutated is located on the X chromosome. As with Duchenne and Becker muscular dystrophies, boys are much more likely to inherit this disease than girls. The two other forms of the diseases are equally common in boys and girls.

The form of the disease caused by a mutation on the X chromosome is usually diagnosed when a person is in his or her 20s or 30s. Researchers have narrowed down where on the X chromosome the gene causing this disease is located but have not yet identified the gene itself. However, there is another gene in the same region of the X chromosome that is sometimes mutated in people with Emery-Dreifuss muscular dystrophy. Researchers are still doing experiments to make sure that this gene, which makes a protein called emerin, is the same gene that they have identified through other experiments. The emerin protein is usually found embedded in the membrane that surrounds the nucleus of smooth, heart, and skeletal muscle cells. However, researchers do not know what role emerin plays in the body or how mutations in emerin lead to muscle weakness.

The two forms of Emery-Dreifuss muscular dystrophy that are not caused by genes on the X chromosome appear very much like the X-linked form. Although neither of the genes has been identified, researchers do know that mutations in one gene cause the disease in a recessive fashion while mutations in the other gene cause the disease in a dominant fashion.

IDENTIFYING THE DISEASE

Emery-Dreifuss muscular dystrophy is the only muscular dystrophy that starts out in the upper arms, causes early contractures, and almost always includes cardiomyopathy as an additional symptom. People with this form of muscular dystrophy do not usually have enlarged calves, which distinguishes this form from Duchenne, Becker, or limb girdle muscular dystrophies. Like other forms of muscular dystrophy, there is no treatment for the disease. However, because people with Emery-Dreifuss muscular dystrophy also have cardiomyopathy with a high risk of sudden death, it is important for these people to be prescribed drugs or get a pacemaker inserted to help the heart function.

Once a male has been diagnosed with Emery-Dreifuss muscular dystrophy, doctors will often examine the heart of any woman in the family who might be a carrier for the disease. (See Chapter 8 for X-linked inheritance.)

Most carriers do not show any signs of muscle weakness, but they can have a mild form of cardiomyopathy that poses some risk of sudden death.

Oculopharyngeal Muscular Dystrophy

Oculopharyngeal muscular dystrophy (OPMD) starts out causing muscles to get weaker in the eyes ("oculo") and throat ("pharyngeal"). The disease usually does not show symptoms until a person is in his or her 50s. It usually spreads slowly to the thighs and hamstrings, then to muscles throughout the body. Over time, the muscles can get so weak that a person has to use a wheelchair to get around.

Early in the disease, a person may have a hard time swallowing food or holding the eyes all the way open. As the disease gets worse, it sometimes also makes the tongue weak.

GENETICS

OPMD can be inherited in either a dominant or recessive fashion. The mutations that cause both forms of the disease are in the same gene, which makes a protein called polyadenylation binding protein 2. The mutation in this gene causes what amounts to a genetic stutter. In a gene, each three base pairs of DNA describe one of the building blocks of a protein. This mutation causes the same protein building block to be inserted several times in a row before continuing with the rest of the protein. More repetitions cause the OPMD to have an earlier onset and be more severe.

People with the mutation causing OPMD usually come from a French-Canadian background or are Bukhara Jews living in Israel. Researchers think that the people who founded those populations probably had the mutation, then passed it on to their children. Because these populations are small and tend to intermarry, the mutation remained common in those populations. OPMD is quite rare outside the French-Canadian or Bukhara Jew population.

One hallmark of OPMD is that the nuclei of muscle cells contain stringy filaments. It turns out that when the protein contains a long series of repeated building blocks, it cannot perform its normal function. Instead, the protein folds up into an elongated shape, and these folded proteins bind to each other to form long chains. Researchers think the filaments in the nucleus are from chains of folded polyadenylation binding protein 2. Researchers still do not know whether it is these filaments interfering with the nucleus that causes the muscle weakness or the fact that the protein is not doing its normal job in the cell.

IDENTIFYING THE DISEASE

Doctors can usually diagnose OPMD based on the fact that the disease starts late in life and usually first shows symptoms in the face. However, if there is any uncertainty a doctor can do a muscle biopsy. The stringy fibers found in the muscle nucleus of people with OPMD aren't found in any other type of muscular dystrophy. Researchers can also look for a mutation in the gene that causes OPMD. This expansion is easier to identify than other types of mutation, so this test is now available at many commercial genetic testing laboratories.

Congenital Muscular Dystrophy

Congenital muscular dystrophy has several different names, including myotonia congenital, amyotonia congenital, congenital myositis, arthrogryposis multiplex, or rigid spine syndrome. The many different names reflect the many different forms congenital muscular dystrophy can take. Babies are usually born with the disease or start showing signs of muscle weakness very early in life. Contractures usually start when the child is still very young or may already have started before the child is born. Unlike most forms of muscular dystrophy, children born with the congenital form of the disease often have some mental retardation.

Sometimes babies with congenital muscular dystrophy move less while they are still in the womb. Children may have problems feeding because the muscles in their face are weak. Unlike other forms of muscular dystrophy that affect specific groups of muscles, children with the congenital form of the disease are weak throughout their entire body.

Children usually stay weak throughout their lives, but the muscles do not usually degenerate. Babies can usually sit up on their own by about three years, but usually cannot stand or walk without support. How long a child with congenital muscular dystrophy lives depends on how well the child is cared for. The most common cause of death is from not breathing well at night. If parents recognize the signs, such as drowsiness or headaches, they can have the child sleep with a device that keeps air flowing into the mouth. Walkers, wheelchairs, and other devices can help kids lead fairly normal lives. Some people can live into their 20s or 30s with the disease.

GENETICS

The congenital muscular dystrophies can take on many different forms in part because they seem to be caused by as many as eleven different genes. Some of these genes have been identified, but many have not. Instead, researchers assume they are caused by different genes because the disease follows a unique pattern in one family. It is possible that, as with Duchenne and Becker muscular dystrophies, two different forms of congenital muscular dystrophies are actually caused by mutations in the same gene.

About half of all babies born with congenital muscular dystrophy have a mutation in a gene called laminin-$\alpha 2$. The protein made by this gene forms a long filament that is part of the network of proteins on the muscle fiber membrane. It is part of the same network as dystrophin and other proteins that are also involved in some forms of muscular dystrophy. This protein is also found in some nerves, which could explain why many children with congenital muscular dystrophies also have learning problems.

Another form of the disease, often called Fukuyama congenital muscular dystrophy, is rare worldwide but is the second most common form of muscular dystrophy in Japan. Most kids with this form of the disease are not able to walk and are extremely mentally retarded. Kids with Fukuyama muscular dystrophy usually die in their teens.

Researchers have found that a gene making a protein called fukutin is mutated in people with this form of the disease. Although researchers do not know what role fukutin plays in the cell, they have found that laminin-α2 is also disrupted in these people. It is possible that fukutin somehow interacts with laminin-α2 in the cell.

Many other forms of congenital muscular dystrophy exist, with varying degrees of weakness, mental retardation, and contractures. So far researchers have not been able to learn how these forms of the disease are inherited.

IDENTIFYING THE DISEASE

Doctors can usually identify congenital muscular dystrophies based on how early the child starts showing muscle weakness and contractures. However, they can sometimes be confused with either Duchenne or Becker muscular dystrophies. One difference is that the congenital muscular dystrophies tend to stay relatively stable, while children with the Duchenne or Becker forms of the disease get progressively worse.

When doctors suspect congenital muscular dystrophy, they often take a muscle biopsy and look for laminin-α2 protein. In most forms of the disease, this protein is either absent or not located in its normal pattern within the cell.

MYOTONIC DISORDERS

All of the myotonic disorders cause the muscles to become tight or stiff ("myo" means "muscle" and "tonic" infers a sustained contraction). People with the disease may also have a hard time relaxing a contracted muscle, such as letting go of something held in the hand. The myotonia is caused by problems with how the muscle fiber conducts the signal to contract. Usually, when the signal sweeps down the muscle fiber from the nerve, some molecules seep into the muscle and others escape. After the contraction, the muscle actively moves these molecules—primarily sodium, potassium, and chloride—back to the appropriate side of the membrane.

In people with myotonic disorders, the pores that open to let these molecules through during a signal to contract are leaky and let the molecules through even when there is no signal to contract. The muscles interpret the presence of these molecules as a contraction signal, letting calcium into the muscle and contracting all the time rather than specifically when instructed to do so by a nerve.

Myotonic disorders can be caused by mutations in the proteins that make any of the pores allowing sodium, potassium, or chloride into or out of the cell. It can also be caused by alterations in the pores that allow calcium into the muscle, giving the ultimate signal to contract. In some forms of the disease, a person can only use the muscle a few times before it becomes too stiff to contract again. In another, almost opposite form, the muscles are very stiff after sitting still for a long time, but start to function normally after a few minutes. The diseases can begin at any age, depending on the mutation. The most common form of myotonic disorder is myotonic dystrophy.

In myotonic dystrophy, the muscles become stiff and are constantly contracted. It affects skeletal muscles all over the body and can also affect the smooth muscles and heart muscle. People usually first show signs of the disease as an adult, when it becomes difficult to hold the eyes open or to release the grip. People with the disease also tend to develop cataracts, even in relatively mild forms of the disease.

Symptoms of myotonic dystrophy don't usually start until a person is in his or her 40s or 50s. Because of this older age, people sometimes attribute their contracted muscles to general stiffness or to arthritis. The disease is often not diagnosed until a person has problems releasing things held in the hand. Women are sometimes diagnosed because their inability to hold the eyes open makes them look unattractive in pictures. The muscles get progressively weaker, especially the muscles of the arms, legs, and face.

Unlike most diseases of the skeletal muscles, the smooth muscles and heart muscles are sometimes affected in myotonic dystrophy. Smooth muscles regulate some aspect of almost all organs. When they do not function normally, they can cause a wide range of problems that are a common cause of death in people with myotonic dystrophy. The most common smooth muscle problem is with swallowing, along with stomach pain and irregular bowel movements. People with myotonic dystrophy often have heart problems, including cardiomyopathy or an irregular heartbeat. Doctors will use a pacemaker to make sure the heart beats properly in people with the disease and to prevent sudden death.

Although the most common form of myotonic dystrophy does not begin until adulthood, a congenital form of the disease also exists. Children born with the disease have a hard time eating and swallowing. As they get older, they have extremely weak face and jaw muscles, which cause the face to be slack with the jaw usually open. Kids with the disease often have severe mental retardation, in part from the disease itself and in part because of brain damage from not being able to breathe well when young.

Genetics

Myotonic dystrophy is caused by a mutation in a gene that makes a protein called myotonic dystrophy protein kinase (DMPK). As with OPMD, the mutation in DMPK is actually a genetic repeat. A small region of the gene that codes for a single building block in the protein gets repeated. The number of repeats tends to increase in each generation and also increases how severe the disease turns out to be. Sometimes, when doctors diagnose a new case of myotonic dystrophy in a family with no history of the disease, they may notice that one of the parents has very mild symptoms that would not otherwise have been noticed and that other family members have cataracts as the only sign of the disease. This situation, where parents show mild symptoms and children show much more severe symptoms, is called disease anticipation.

Once the number of repeats has gone up enough that it causes the disease, that person will pass on the mutated gene to some children, who will also develop the disease. The number of repeats accounts for the difference between a typical case of myotonic dystrophy and a congenital case. In a typical case that begins in adulthood, the person has only a few repeats. Children born with the disease have a very high number of repeats. Although researchers know quite a bit about how mutations in DMPK are involved in the disease, they do not know much about the protein itself.

Identifying the Disease

Weakness in the face distinguishes myotonic dystrophy from most forms of muscular dystrophy. At the same time, the fact that some of the muscles in the body become tense also distinguishes the disease from OPMD, which can look similar when a person is first diagnosed. It is important for doctors to correctly diagnose this disease so they can take the appropriate steps to treat any possible cardiomyopathy.

Congenital myotonic dystrophy can appear a lot like congenital muscular dystrophy to doctors. The main difference is that children with myotonic dystrophy have much more severe weakness in the face. In most cases, the best way to conclusively diagnose the disease is to look for the genetic repeat in the DMPK gene. This test can confirm a diagnosis and is also useful for prenatal testing.

MYOPATHIES

Congenital Myopathies

A myopathy is any form of disease that affects the muscle. Cardiomyopathy, for example, is a muscle disease that affects the heart. Congenital myopathies are a group of diseases that all cause babies to be born with diseased muscle. To some extent, a congenital muscular dystrophy is also a type of myopathy, although the dystrophies are more specifically a disease in which the muscles waste away.

The congenital myopathies usually cause mild muscle weakness that does not get worse as a child grows older. They can be caused by mutations in a wide range of genes, most of which make proteins that are involved in muscle contraction or in the membrane surrounding the muscle fiber. The individual diseases are so rare that it has been hard for researchers to learn much about how these mutations cause myopathy.

Inflammatory Myopathies

Some forms of muscle disease happen when the immune cells invade the muscles, causing them to become inflamed. Many different forms of **inflammatory myopathy** exist, which can start anywhere from early childhood to adulthood, depending on the disease. For any of the diseases, the muscles become progressively weaker as the immune cells invade and damage the muscle fibers.

CAUSES

In most cases, the inflammatory myopathies are a form of **autoimmune disease**. The immune cells inappropriately identify proteins in the muscle as foreign and attack those proteins to prevent them from hurting the body. This type of reaction is effective against invaders such as bacteria or virus, but they can be completely disabling when directed against the body. Arthritis, multiple sclerosis, and juvenile diabetes are all forms of autoimmune disease. Depending on which muscle protein the immune cells have targeted, the disease will primarily affect different muscle groups.

Some researchers have suggested that being infected by a virus is the first step in developing an inflammatory myopathy. The immune cells initially fight off a virus that is in the muscle. Over time, the immune cells begin attacking the muscle itself rather than the virus.

TREATMENT

Inflammatory myopathies cannot be cured, but doctors can often treat the symptoms to help people lead more normal lives. The first step to treating these diseases is distinguishing them from other forms of muscle disease. Diagnosing these myopathies usually requires taking a biopsy from the muscles that are already becoming weak. In an inflammatory disease, this biopsy sample should show inflammation and include an unusually large number of immune cells.

The most common treatment for inflammatory myopathy is a steroid drug called prednisone, which blocks the immune system. This is the same drug that helps people who have a severe poison oak or poison ivy rash—both of which are caused by an overzealous immune reaction. Although prednisone may initially keep the immune response in check, it does have serious side effects if it is taken for a long time. These include risks of heart disease, some forms of cancer, and making a person susceptible to new infections. Doctors will usually cut a person's dose so that it is as low as possible and still minimally effective. After a while, most people stop responding to prednisone and need to move on to stronger types of steroids, all of which can have serious side effects including an increased risk of cancer and heart disease, weight gain, and infection.

Distal Myopathies

There are several different forms of **distal myopathies**, all of which cause the muscles farthest from the center of the body (distal) to become weak. Most myopathies start with muscles close to the center of the body (proximal) such as the hamstrings, thighs, shoulders, and upper arms. Distal myopathies are sometimes called Welander myopathy after a Scandinavian study in 1951 that first verified the disease.

Distal myopathies are rare and are sometimes classified into three forms. Two of these forms don't show signs until adulthood, while another form begins during childhood. In all forms, the disease usually starts out in the wrists and fingers or ankles and toes and then begins affecting other muscle groups.

Distal myopathies can be extremely variable. One form of distal myopathy—most commonly referred to as Welander myopathy—doesn't usually begin until a person is in his or her 50s or as late as in his or her 70s. This form progresses very slowly and doesn't usually shorten a person's life. Other forms start when a person is in his or her 20s or 30s and can progress so quickly that a person can end up in a wheelchair within 10 to 15 years. However, in one form of the disease called Miyoshi myopathy, it is the calf muscles that are most severely affected. A person usually first notices that they can't walk on their toes or climb stairs. Over time, the calf muscles waste away to almost nothing.

GENETICS

Although researchers have broken the distal myopathies into three different forms, they can be caused by mutations in about nine different genes. The diseases can be inherited in either a dominant or recessive fashion, depending on the mutation.

In some cases, the mutation that causes a distal myopathy is in the same gene as one that can cause limb girdle muscular dystrophy. For example, Myoshi myopathy, is a type of adult-onset distal myopathy, caused by a mutation in a gene called dysferlin, which can also be mutated in limb girdle muscular dystrophy. Researchers suspect that another distal myopathy is also caused by a mutation in the dysferlin gene, but this form of the disease doesn't become obvious until a person is in his or her 50s. Because these mutations are so rare, researchers don't understand why one mutation in the dysferlin gene would cause muscles of the back, shoulders, and chest to become weak, while other mutations would cause the calves and lower arms to become weak.

Very few other genes have been identified that cause distal myopathies. In some cases researchers have narrowed down a small chromosomal region where the gene is likely to be located, but so far they have not identified a specific gene.

IDENTIFYING THE DISEASE

The distal myopathies are not usually confused with other forms of muscular dystrophy. None of the muscular dystrophies begins in the hands, feet, or calves, or begins as late in life as some forms of distal myopathy.

METABOLIC DISEASES

The ability to make ATP from sugars, fats, or proteins requires enzymes in the main part of the cell, enzymes in the mitochondria, and proteins in the mitochondrial membrane that make up the electron transport chain. When any of the genes making these essential proteins is mutated, the cell cannot make enough ATP. These mutations make it hard for all cells of the body to get enough energy, but because the muscles require the most ATP the skeletal muscles and heart are what usually show the first symptoms of these diseases.

People with any form of metabolic disease can have a range of symptoms including being extremely tired from even mild exercise, severe muscle cramps, and progressive muscle weakness. These symptoms usually become obvious when a person is still a child, but might not get diagnosed until early adulthood.

Some of the proteins involved in these diseases play an active role in glycolysis, the citric acid cycle, or oxidative phosphorylation. Other proteins are involved in breaking down glycogen into glucose in either the liver or in the muscle, or in preventing glycogen from getting stored inappropriately in cellular compartments called lysosomes. The diseases may also result in diabetes, because the muscles may be unable to take up glucose in order to use it for energy.

Genetics

All of the metabolic diseases are inherited in a recessive fashion. A person must inherit two mutated copies of a gene in order to develop the disease. If a person has at least one functional copy of a metabolic gene, they will make some protein and the cell will be able to generate ATP.

Chapter 2 stated that mitochondria also have DNA and contain genes that make proteins involved in the citric acid cycle and in the electron transport chain. These genes can also be mutated in metabolic diseases. These mutations are always inherited from the mother, because all mitochondria come from the mother's egg. Mitochondrial mutations often make the heart muscle weak in addition to the skeletal muscles. Chapter 10 has information about diseases of the heart muscle caused by mitochondrial mutations.

Identifying the Disease

The initial symptoms of being tired with extremely mild exercise and muscle cramps or pain may suggest a metabolic disorder. Verifying that diagnosis usually involves testing the forearm muscles by having a person hold an object or squeeze a ball. Doctors can measure the amount of lactate in the blood or in a muscle biopsy after that exercise to see if it is normal. It is also possible to take muscle cells from that biopsy and grow them up in a lab dish to understand more about what proteins are not functioning properly in the muscle.

Treating the Disease

For the metabolic diseases, it is important to understand exactly which form of disease the person has. In some cases, people can modify their diet so that they eat more of the foods that their body can easily use for energy, such as eating more carbohydrates or eating more short-chain fats rather than long-chain fats. For another form of the disease, it is important for people to get some mild exercise to maintain their aerobic endurance. These people have very limited ability to make ATP anaerobically. In other people, who have defects in the mitochondrial proteins, they need a steady supply of glucose to fuel anaerobic respiration, because they can generate very little energy through aerobic respiration.

10

Diseases of the Heart Muscle

Heart disease is the leading cause of non-accident-related death in the United States. Of the various types of heart disease, some have to do with the heart muscle itself whereas others have to do with blockages in the arteries that carry blood to the body. The most common form of heart disease is coronary artery disease (also known as CAD), in which the arteries that feed oxygen to the heart muscle become clogged. Eventually, these roadblocks completely prevent the heart muscle from getting oxygen and it can no longer contract properly, causing a heart attack. Although CAD and heart attacks are major health concerns around the world, the diseases themselves are not specifically diseases of the heart muscle.

This chapter will focus on a class of diseases called cardiomyopathies ("cardio" means heart, "myo" means muscle, and "pathy" means disease) that specifically affect the heart muscle rather than the arteries leaving the heart. Although cardiomyopathy affects relatively few people compared to the millions who have all forms of heart disease, cardiomyopathy is the most common reason for needing a heart transplant and is the leading cause of sudden death in young athletes.

TYPES OF CARDIOMYOPATHY

Cardiomyopathies can take many different forms, but in each case the heart muscle becomes weak and unable to contract properly. The weakened heart can't pump with enough force to send blood throughout the body where it is needed. Some forms of cardiomyopathy severely limit a person's activity, whereas other forms go unnoticed until the heart is pushed by strenuous exercise.

The different types of cardiomyopathy stem from how the diseases originate. Some are caused by inheriting mutations in genes that make heart muscle proteins. Other cardiomyopathies come from infections, from drinking too much alcohol, or from unknown causes. Whatever the underlying reason for the disease, because the heart muscle is not able to divide and repair itself and does not have the pool of satellite cells that repair damage in skeletal muscle, cardiomyopathies get progressively worse until the heart is completely unable to pump blood. At this point the only treatment is a heart transplant. How quickly the disease gets worse depends both on the type of cardiomyopathy and on the individual person.

Doctors usually describe cardiomyopathy as being either ischemic or nonischemic. Ischemic cardiomyopathy refers to damage to the heart muscle caused by a heart attack. The heart muscle fibers that had their oxygen supply disrupted by the blocked artery die and cannot regenerate. This type of cardiomyopathy does affect the heart muscle, but the blocked artery rather than the muscle itself is the underlying reason for the disease. As with other forms of cardiomyopathy, once the heart muscle is damaged by a heart attack it gets progressively weaker and unable to pump blood. This chapter will not focus on ischemic cardiomyopathy, although it is a very important disease in health care.

Nonischemic cardiomyopathy is any type of cardiomyopathy that is not caused by a heart attack. This form of the disease can be further broken down into **hypertrophic cardiomyopathy**, **dilated** (or congestive) **cardiomyopathy**, and **restrictive cardiomyopathy**.

HYPERTROPHIC CARDIOMYOPATHY

Hypertrophic cardiomyopathy is the second most common type of cardiomyopathy, affecting about 300,000 people in the United States. About one in 500 people are born with the disorder. The disease is sometimes known by other names such as idiopathic hypertrophic subaortic stenosis (IHSS), asymmetrical septal hypertrophy (ASH), or hypertrophic obstructive cardiomyopathy (HOCM).

By whichever name you call the disease, the heart muscle fibers in people with hypertrophic cardiomyopathy develop abnormally, with muscle fibers forming a tangle rather than the regularly branching fibers in normal heart muscle. The weaker muscles cause the walls of the heart to compensate by growing thicker in order to pump enough blood. The muscle is particularly thick in the region that divides the lower left and right sides of the heart, called the **septum**. This thicker septum pushes in on the left region of the heart called the left **ventricle**. This is the chamber that is mainly responsible for pumping blood throughout the body. (The Circulatory System volume in this series has more information about the anatomy of the heart.) That wall is usually thinner than 12 millimeters, but it is usually more than 15 millimeters thick in a person with hypertrophic cardiomyopathy.

Although most people with the disease have most of the thicker muscle between the left and right ventricles, about 25 percent of people have heart muscle that is evenly thicker around the entire left ventricle (called concentric hypertrophy). In about 10 percent of people with hypertrophic cardiomyopathy, the muscle is almost exclusively thicker near the bottom portion of the left ventricle (called apical hypertrophy) shown in Figure 10.1. This pattern is particularly common in Japan.

The thicker wall means that the ventricle can hold less blood. What's more, the abnormal muscle fibers can't relax properly to let blood into the already smaller chamber. Another problem arises when the muscle thickening is asymmetrical with most of the hypertrophy near the top of the septum shown in Figure 10.2. With this pattern, the septum can block the opening where the blood leaves the heart. These problems all add up to a heart that cannot effectively pump blood throughout the body.

Hypertrophic cardiomyopathy usually becomes apparent when a child goes through puberty and all the muscles in the body are growing very quickly, although some forms of the disease do not cause symptoms until a person is his or her 50s or 60s. Once the muscle goes through its initial growth spurt, it usually stays about the same thickness throughout a person's life.

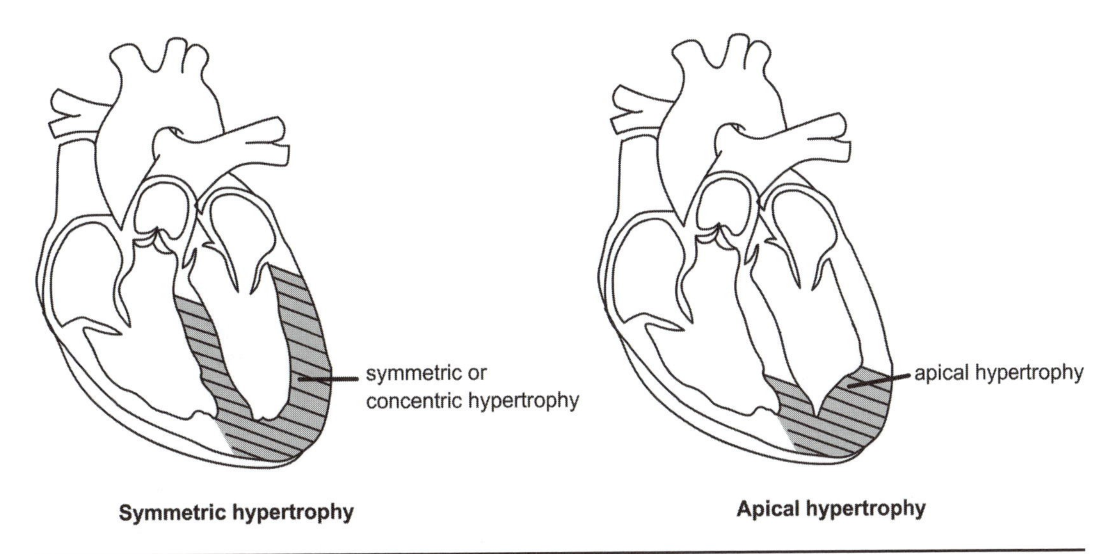

Figure 10.1. Symmetrical and apical hypertrophy.

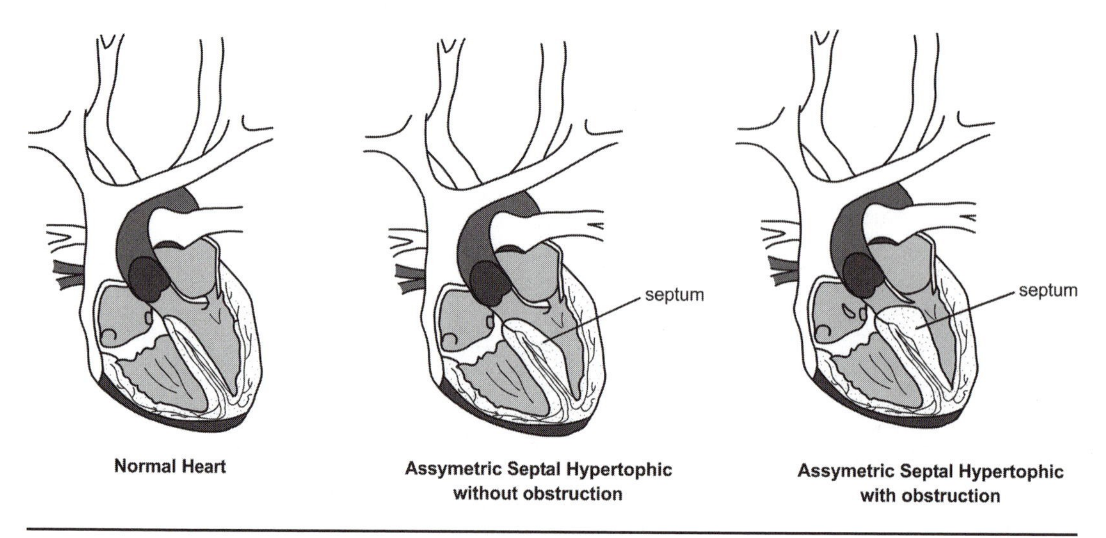

Figure 10.2. Hypertrophic cardiomyopathy with and without obstruction.

Symptoms

Some people with hypertrophic cardiomyopathy have no symptoms of the disease. In fact, some people live symptom-free until one day the heart is no longer able to maintain a regular heartbeat. Instead, the heart quivers but doesn't deliver blood to the body. This is the most common cause of sudden death during intense athletics in otherwise healthy athletes. About 2–3 percent of people with hypertrophic cardiomyopathy die from sudden death each year. Because sudden death can happen in a person with no other symptoms of the disease, people who have even very early stages of hypertrophic cardiomyopathy are usually banned from playing competitive sports.

Other people may have no symptoms from their enlarged heart, then notice that they become more out of breath than usual during exercise. This mild limitation may not trigger a visit to the doctor, but as the heart gets weaker, a person may become out of breath during day-to-day activities such as climbing stairs or cleaning. The heart may be strong enough to pump blood when relaxing, but when the muscles need more oxygen the heart can't pump enough blood to the muscles to support ATP production. People may also faint during exercise from the lack of oxygen, or they may notice that their heart beats very hard and quickly. None of these symptoms is unique to hypertrophic cardiomyopathy, but they do indicate that the heart is not functioning normally.

Sometimes a person with hypertrophic cardiomyopathy may notice that his or her chest hurts. This is probably because the weak heart is not pumping enough blood to the heart muscle itself. Although most of the heart may receive enough oxygen, the greatly enlarged muscle between the lower left and right sides of the heart often cannot get enough oxygen and begins to hurt. The pain usually happens during exercise but can also happen when relaxing later in the disease.

Another symptom of hypertrophic cardiomyopathy is an irregular heartbeat. The enlarged muscle can't always conduct the signal to contract to other muscle fibers in the heart. When this happens, the heart may skip a beat or add extra beats and may begin to beat very quickly. This often happens in conjunction with feeling lightheaded or sweating. About 10 to 20 percent of people with hypertrophic cardiomyopathy may also go on to develop dilated cardiomyopathy.

In the later stages of hypertrophic cardiomyopathy, a person may end up having **heart failure**, when the heart is simply not able to pump enough blood to the body, even when resting. Heart failure can be caused by many different forms of heart disease. When the heart can't pump enough blood, people tend to accumulate fluid in the lungs. Because of this congestion, heart failure is often called **congestive heart failure**. Heart failure is relatively uncommon in hypertrophic cardiomyopathy and usually only occurs when a person has also gone on to develop dilated cardiomyopathy.

Causes

INHERITANCE

Researchers have known since the 1950s that hypertrophic cardiomyopathy tends to run in families and is almost always caused by inheriting a mutated gene. The disease is inherited in a dominant pattern, which means that a person only has to inherit a mutated gene from one parent in order to develop the disease as shown in Figure 10.3. For some of the mutated genes, inheriting two mutated copies makes the heart so weak that the embryo would likely die while the heart is still developing. For other genes, inheriting two copies causes dilated cardiomyopathy instead of hypertrophy.

Depending on the mutated gene, the disease may start earlier or later in life, may be more or less severe, and may cause different patterns of hypertrophy in the heart. Some gene mutations also put people at higher risk for sudden death during exercise than other mutations.

This variation in how the disease progresses is not just due to which gene is mutated. When researchers look at members of a family who have all inherited the same mutation, they still see quite a bit of variation between family members. They think that the mutation itself is only part of the equation that determines exactly how that person's hypertrophic cardiomyopathy develops. Other factors such as whether a person exercises, body weight,

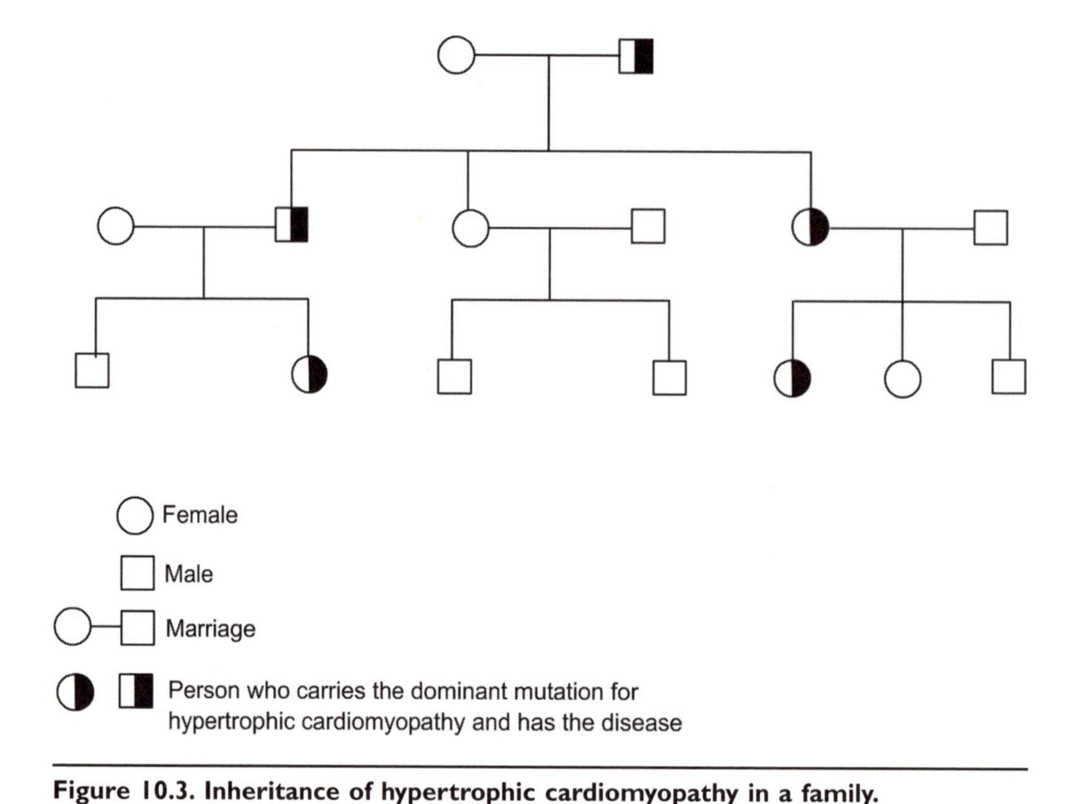

Figure 10.3. Inheritance of hypertrophic cardiomyopathy in a family.

and diet all play a role in determining how severe the cardiomyopathy turns out to be.

Another finding that supports the idea that the gene mutation is only one factor that determines the extent of the hypertrophy comes from mice. Researchers have created mouse strains that carry the same mutation that causes hypertrophic cardiomyopathy in humans. When this gene is mutated in one strain of mice, all mice with one mutated copy of the gene have hypertrophy in the left ventricle. However, when the gene is mutated in a different strain of mouse, only half of the mice end up having hypertrophic cardiomyopathy. Researchers suspect that other genes that are different between the two mouse strains account for this difference in disease risk.

Researchers are studying mouse strains that inherit a risk for hypertrophic cardiomyopathy to learn more about how to prevent the disease. For example, they may learn that mice getting certain vitamins or food supplements have a lower risk of developing the disease or show symptoms much later than other mice with the same mutation. This type of research could help people who have a family history of the disease lower their own disease risk.

MUTATED GENES

So far, researchers have found twelve genes that may be mutated in people with hypertrophic cardiomyopathy. Of these, ten make proteins that are involved in how the sarcomere contracts, such as a heart-specific version of myosin, troponin, tropomyosin, vinculin, and others. Two other genes make proteins that are not involved in the sarcomere. One of these makes a protein that senses how much ATP is in the cell. Researchers think mutations in this gene cause hypertrophic cardiomyopathy when the cell cannot sense ATP levels and so doesn't make enough ATP. The muscle fibers then do not have enough energy to contract.

Researchers think that by having sarcomeres that cannot contract properly, the muscle surrounding the left ventricle grows larger to compensate for its weakness. However, despite decades of research in this area, researchers have very little idea why the mutations cause the muscle fibers to become disoriented. They also do not know the exact signals that determine whether the muscle becomes thicker around the entire left ventricle or only in the septum.

One thought has been that regardless of why the muscle fibers are weak, it is that weakness itself that starts a series of reactions that leads to hypertrophy. This would mean that the hypertrophy would follow the same path regardless of what gene is mutated. In this model, other factors such as diet, weight, or environmental factors would then influence exactly how that hypertrophic pathway proceeds, such as when symptoms start or the pattern of hypertrophy around the left ventricle.

CALCIUM REGULATION

Heart muscle fibers in mice with hypertrophic cardiomyopathy are extremely sensitive to calcium. Researchers have also found that these mice have lower calcium reserves in the heart muscle itself than other mice. This sensitivity to calcium explains why the heart muscle does not relax properly in people with hypertrophic cardiomyopathy. If small amounts of calcium are all that is needed in order for the muscle to contract, the residual calcium in the fiber after a contraction may still be enough for the heart muscle to remain contracted rather than relax normally to allow the heart to fill with blood.

Some of the mutations that cause hypertrophic cardiomyopathy are in troponin or other proteins that do interact directly or indirectly with calcium. These mutations may make it easier for calcium to bind troponin and cause the fiber to contract in response to lower overall calcium levels. However, other mutations that cause the disease have no impact of calcium-binding proteins.

Although researchers still haven't put all the pieces together to understand how calcium signaling may be disrupted in hypertrophic cardiomyopathy, this is an active area of research. As part of this research, doctors are learning that some drugs, such as cyclosporin used to prevent transplant rejection, also affect calcium signaling and may be more toxic to people with hypertrophic cardiomyopathy.

Diagnosis

It used to be that hypertrophic cardiomyopathy was not diagnosed until a person reported heart-related problems. Now that doctors know that the disease runs in families, they are more likely to screen family members of a person with the disease even if those family members do not yet have symptoms. By doing this, more people with the disease are being diagnosed in their teens or 20s. Current recommendations are that a person with a family history of hypertrophic cardiomyopathy be screened every year after the age of 12, when symptoms usually first appear, until age 25. If the heart has not enlarged by that age, the person is unlikely to have the disease, though doctors still recommend screening every five years in case the heart enlarges late in that person.

If doctors do find young people with hypertrophy, they recommend that those people avoid strenuous exercise in case of sudden death. Doctors also recommend a heart-healthy lifestyle such as not smoking, eating a healthy diet, and not drinking in excess so that the heart muscle has the best possible chance of staying strong.

Doctors have several tools for distinguishing between different forms of heart disease. Listening to the heart with a stethoscope can tell a doctor whether or not the heart is beating regularly. This may not pick up hypertrophic cardiomyopathy if the disease has not progressed far enough that the heartbeat is affected. The stethoscope may also pick up a murmur that happens in about 30 to 40 percent of patients when the opening from the left ventricle is partially blocked and the blood circulates around the cavity instead of flowing out of the heart. This murmur only happens in those people whose heart muscle has become so thick that the blood flow is obstructed, so the absence of a murmur does not rule out hypertrophic cardiomyopathy.

The best way to detect hypertrophic cardiomyopathy is using **echocardiography**. In this test, a doctor bounces sound waves off of the heart. Some of the sound waves bounce off of the outside of the heart, while others pass through the outer heart layers and bounce off of the inner wall of the heart and back to the recording equipment. The way those sound waves bounce back forms a complicated image that a doctor can interpret to determine the size and shape of the heart and the structures within the heart. In hypertrophic cardiomyopathy, the thicker heart walls show up clearly using this test.

The echocardiogram can easily detect an enlarged heart, but it may give a confusing result in athletic people whose heart muscles are larger than usual because of regular exercise. In athletes, the larger heart contains healthy muscle that contracts with a lot of force to pump blood throughout the body. The muscle in a person with hypertrophic cardiomyopathy may be about the same size as in an athletic person, but the tangled muscle fibers don't contract properly or with much force. One way for doctors to tell the difference between an athletic heart and one with hypertrophic cardiomyopathy is that the diseased heart muscle is mainly thicker around the left ventricle, whereas the athletic heart muscle is thicker throughout the heart.

Doctors may also use a **electrocardiogram (EKG)** to record how electricity moves throughout the heart. The disorganized fibers don't conduct electricity normally and will give an unusual reading during this test. However, other heart conditions also cause an abnormal electrocardiogram, so this test is not always conclusive.

If none of these tests makes clear what type of heart disease a person has, a heart muscle biopsy can reveal whether or not the heart muscle looks normal. In a person with hypertrophic cardiomyopathy, the heart muscle fibers are visibly tangled under a microscope compared with normal heart muscle. Doctors call this condition "myocardial disarray."

Treatment

Doctors have several options for how they may treat hypertrophic cardiomyopathy, depending on how strong the heart is. None of the treatments cures the disease, but they can prevent the heart from becoming weaker. If the disease is detected before any symptoms, a person has the best chance of keeping the heart as strong as possible and delaying the heart's decline.

If the heart has already become weaker when the disease is detected, then some drugs can help make the heart's job easier. These include some drugs that relax the heart, that slow the heartbeat so each beat can be stronger, or that help the heart maintain a steady rhythm. Once a person already has heart failure, then drugs that cut back the amount of water in the blood can make the blood thicker and reduce the amount of fluid that builds up in the lungs.

Some patients eventually need surgery to repair the heart muscle. Although surgery can't correct how the fibers are arranged, the doctor can scrape away some of the thickened muscle so that the inner cavities of the heart can contain more blood. The surgery also removes any muscle that is blocking the exit from the left ventricle. This procedure helps about 70 percent of patients but can cause deadly side effects in about 1–3 percent of people with the disease. Some other people do well after the surgery but have a severely slowed heartbeat. These people end up needing a device called a pacemaker, which attaches to the heart and helps it beat at a constant rhythm. Pacemakers are used to treat many different kinds of heart disease. If none of these other measures works, a person may end up needing a heart transplant.

A relatively new form of treatment avoids the open heart surgery that is needed to remove part of the enlarged heart muscle. In this technique, called alcohol ablation, a doctor injects alcohol into one of the blood vessels that goes to the enlarged portion of the heart. This potent alcohol kills any cells that come in direct contact. The technique is still experimental, but it may be one way to remove excess heart muscle fibers without surgery.

DILATED CARDIOMYOPATHY

Dilated cardiomyopathy is the most common form of nonischemic cardiomyopathy, affecting between 2 and 3 million people in the United States. As the name suggests, in a person with dilated cardiomyopathy the heart grows larger in size as shown in Figure 10.4. The cavities within the heart that hold blood enlarge first, then the heart muscle grows larger in order to pump the extra blood that the heart can now hold. Although the heart enlarges, the muscle is usually not strong enough to pump blood effectively. To compensate, the body often makes more blood, but this extra blood fills the heart even fuller and makes it that much harder for the heart to contract and push the blood out to the body. Dilated cardiomyopathy happens more often in men than in women.

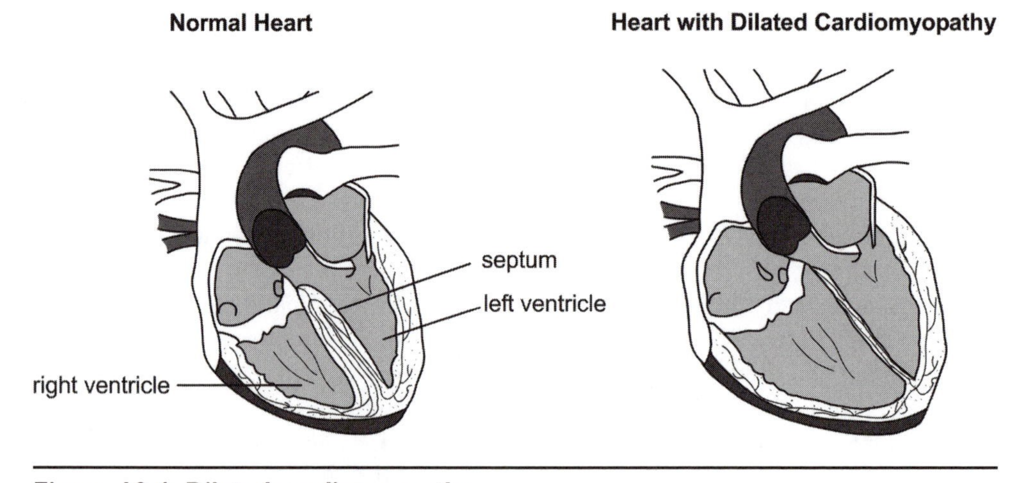

Figure 10.4. Dilated cardiomyopathy.

Symptoms

The heart can start growing larger at any age, but people do not usually notice any signs of the disease until they are in their 40s or 50s. The earliest signs are usually that a person is tired, weak, and short of breath, especially when active. These symptoms happen because the heart is not able to pump blood properly, so other parts of the body including the skeletal muscles do not get enough oxygen. Without oxygen, the muscles cannot make enough ATP. At first a person may only notice being tired during more strenuous activities than usual. At this point the heart is still able to pump blood for everyday activities but cannot increase the heartbeat above that minimal level. Eventually, the heart becomes weak enough that a person gets out of breath even after mild activities like walking up stairs.

If a doctor catches dilated cardiomyopathy at this early stage, it is sometimes possible to keep the heart muscle at a relatively steady state using drugs. But even with these measures, the heart does eventually become weaker. When this happens, fluid tends to build up in a person's lungs and in the legs or feet when lying down. The cough that results from having fluid in the lungs often means that a person with dilated cardiomyopathy seems congested and often gets misdiagnosed as having pneumonia. It is also the reason why dilated cardiomyopathy is sometimes called congestive cardiomyopathy.

In the late stages of the disease, people may have pain in the chest or stomach, or might start having irregular heartbeats. When the heart starts beating irregularly, that means that the muscle is extremely weak and is no longer able to keep contracting. Unfortunately, many people do not get their disease diagnosed until the heart is in this extremely weak state. Because of this late diagnosis, only about half the people who are diagnosed with the disease live another five years, with slightly more women surviving five years than men. Only one quarter of the people live ten years after a dilated cardiomyopathy diagnosis.

At all stages of dilated cardiomyopathy, a person is more likely to develop blood clots than other people. This is because blood moves slowly through the larger heart, giving blood time to coagulate. These clots then leave the heart and go to other tissues where they can block the blood from flowing through a blood vessel. These clots can be deadly if they get lodged in the lungs or brain. Sometimes blood clots are the first sign of dilated cardiomyopathy.

Causes

Although dilated cardiomyopathy is quite common, doctors rarely discover a specific reason for the disease. However, it does seem to happen more often in people who drink excessive amounts of alcohol, especially if the person's diet is not healthy. It can also be caused by high levels of a mineral called cobalt that used to be used in making beer, or by being infected with a virus. Rarely, some women develop dilated cardiomyopathy after giving birth. The condition usually reverses itself but can come back after later pregnancies.

Some drugs that are used to treat cancer can also damage the heart muscle and cause it to enlarge. The most common cancer drug to cause cardiomyopathy is doxorubicin, which is used to treat many forms of cancer including breast cancer and some forms of lymphoma.

VIRAL INFECTION

One particular virus, called coxsackie-B, is particularly prone to infecting the heart muscle and causing dilated cardiomyopathy. When coxsackie-B first infects the heart muscle, some of the fibers die immediately. In other fibers, the virus inserts its genes into the muscle cell's DNA, and the fiber begins to make viral proteins. The immune system recognizes those proteins as foreign and launches an attack.

The problem arises when that attack damages the muscle cells where those viral proteins are being made. When the immune system enters the heart and attempts to fight a virus, it can make the heart inflamed. A person may feel like he or she has a flu and an achy chest that go away within a few days. The initial attack kills some muscle fibers, but an ongoing, lifelong influx of immune cells in the heart is what finally damages the heart to such an extent that dilated cardiomyopathy results.

INHERITANCE

Unlike the hypertrophic cardiomyopathy, dilated cardiomyopathy is rarely caused by inheriting a mutated gene. However, in about 30 percent of people with the disease, another family member will also have the disease. This could be due to an inherited gene, but it could also be because other family members also have heart damage due to drinking excessive alcohol or due to a coxsackie-B viral infection.

In some rare cases, known genetic mutations are the underlying cause of dilated cardiomyopathy. So far, researchers have found about ten genes that, when mutated, increase a person's risk of dilated cardiomyopathy. Most of these genes make proteins that are used to help the heart muscle fiber contract. These include forms of myosin, troponin, tropomyosin, and actin that are only used by heart muscle. Some of these genes are the same ones that can be mutated in hypertrophic cardiomyopathy, although mutations that cause the two diseases tend to happen in different parts of the protein. One idea researchers have is that mutations causing relatively mild changes in sarcomere proteins cause the heart muscle to compensate by becoming hypertrophic. When the mutation seriously disrupts the protein, the heart muscle cannot compensate by enlarging and instead becomes dilated. By this theory, it is the severity of the mutation that determines whether a person develops hypertrophic versus dilated cardiomyopathy.

Like hypertrophic cardiomyopathy, most of the mutations that cause dilated cardiomyopathy are inherited in a dominant fashion—that is, if a person inherits a mutated gene from one parent, the person is at risk of developing the disease. Other forms of dilated cardiomyopathy are recessive, where a person must inherit two mutated copies of the gene in order to have the disease. This pattern of inheritance is shown in Figure 10.5. However, unlike hypertrophic cardiomyopathy, inheriting one of the mutations usually does not ensure that a person will develop dilated cardiomyopathy. It simply increases a person's risk of the disease. Researchers think these gene mutations make a person more susceptible to getting the disease after a viral infection, heavy drinking, or pregnancy.

There seem to be several ways that altered proteins in the heart muscle can lead to dilated cardiomyopathy. Mutations in the myosin gene prevent

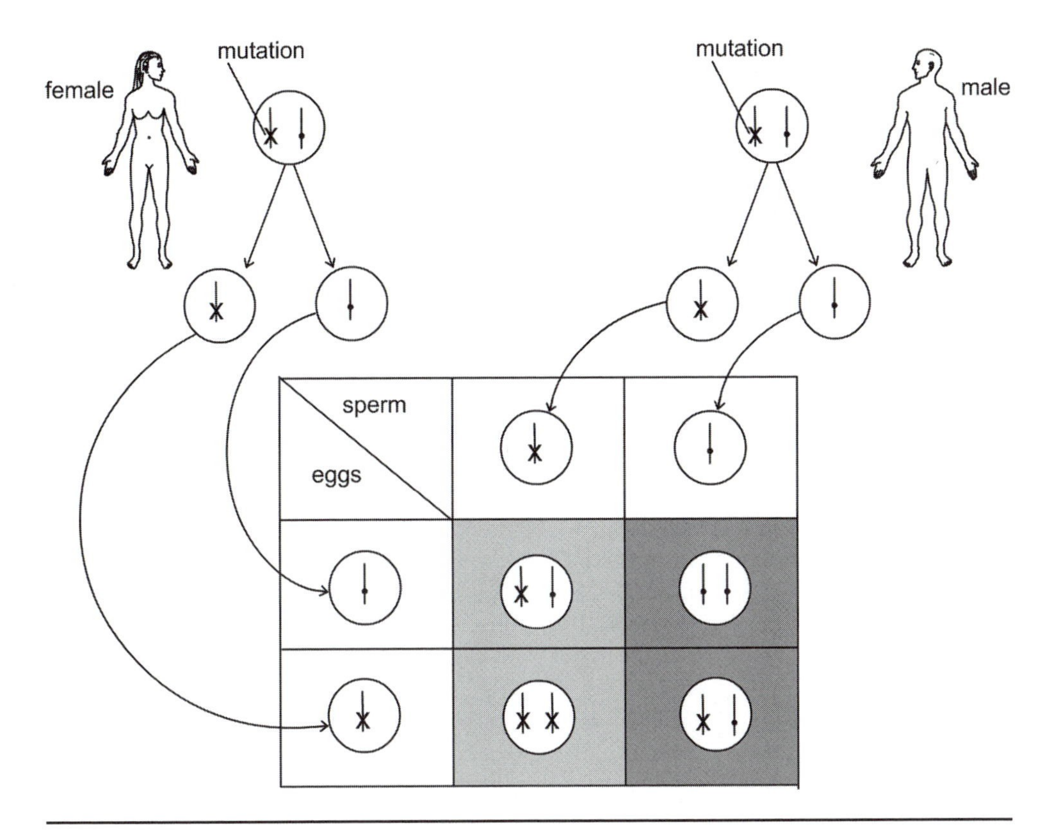

Figure 10.5. Inheriting a recessive disease from two unaffected parents.

the myosin heads from pulling on the actin with enough strength to shorten the sarcomere. Although the fiber receives the signal to contract, it can't contract with the full strength of a normal heart muscle fiber. Mutations in actin and troponin also seem to prevent the sarcomere from contracting normally.

Mutations in proteins called desmin and vinculin seem to cause dilated cardiomyopathy by not transmitting the force of a contraction between muscle fibers. These proteins are part of the matrix that hold a sarcomere together and translate the force of one sarcomere contracting to the entire muscle fiber. When these proteins are damaged, individual sarcomeres may contract but the force generated by those contractions is less than in a normal heart.

Another group of proteins that can be mutated in dilated cardiomyopathy are proteins that assemble on the outside of the sarcomere and stabilize the muscle fiber. These are generally part of the same complex of proteins as dystrophin, which is mutated in muscular dystrophy. (Chapter 8 has more information about how dystrophin mutations cause muscular dystrophy.) In fact, some mutations that cause either Duchenne or Becker muscular dystrophy also cause dilated cardiomyopathy.

In some families, a mutation in dystrophin will cause dilated cardiomyopathy but not muscular dystrophy. There have been too few families with this type of inheritance for researchers to understand why some mutations in this gene affect the heart muscle but not skeletal muscle. However, they think this pattern of inheritance may be due to the dystrophin gene being used differently in skeletal and heart muscle. For some genes, only part of the gene is used to make a protein for one tissue, while other parts of the gene are used to make a protein for another tissue. Researchers think that dilated cardiomyopathy happens when there is a mutation in a part of the dystrophin gene that is used by heart muscle but not skeletal muscle.

As with muscular dystrophies, men are much more likely than women to develop dilated cardiomyopathy due to mutations in dystrophin because the gene is on the X chromosome. (See Chapter 8 for more information about how mutations on the dystrophin gene are inherited.) Men who inherit this form of the dystrophin gene tend to develop signs of dilated cardiomyopathy by the time they are in their teens or 20s and often die within one or two years. Women who carry one copy of this mutation sometimes have no symptoms but may develop cardiomyopathy later in life.

Chapter 2 described how mitochondria generate energy in the form of ATP that muscle cells use to fuel the contraction. It turns out that some of the proteins used in the mitochondria come from genes on chromosomes within the mitochondria themselves rather than from within the cell's nucleus. When these genes have mutations, they prevent the heart from being able to make enough ATP to sustain a regular contraction. This is a particular problem in the heart muscle, which relies exclusively on aerobic respiration in the mitochondria to produce ATP.

Although mitochondria throughout the body may contain a mutation, the heart may be the only tissue that shows any disease because other tissues in the body can compensate with anaerobic respiration. Another reason that mitochondrial mutations cause disease in some tissues but not in others has to do with how cells pass on mitochondria.

Each cell has several mitochondria that came from the mother's egg. Some of these mitochondria may contain a mutation while others do not. Each time a cell divides, the mitochondria also divide, then divvy themselves up at random between the two cells as shown in Figure 10.6. Through this random process, one tissue may end up with a majority of defective mitochondria while other cells have a majority of normal mitochondria. If the heart muscle precursors in the embryo had a preponderance of mitochondria with mutations, then the heart will end up not being able to make enough ATP and can cause dilated cardiomyopathy. The same mitochondrial mutations can also cause diseases in skeletal muscle, or in all muscle types, depending on which tissues have the most defective mitochondria.

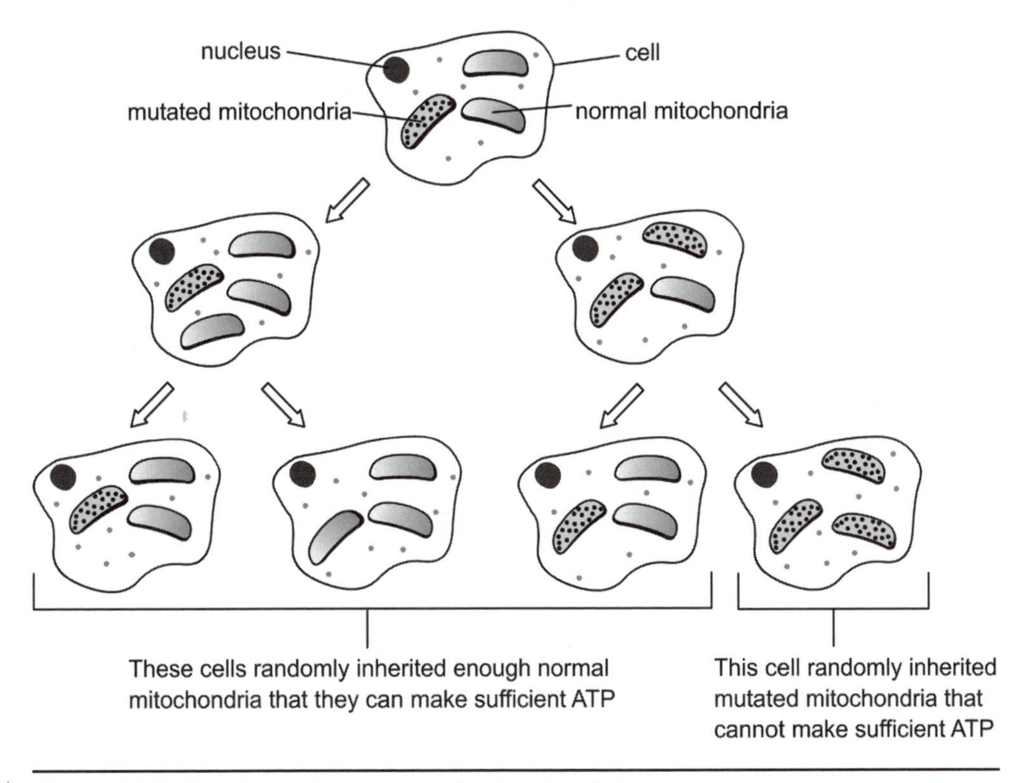

Figure 10.6. Random inheritance of mitochondria.

Diagnosis

Doctors can usually begin to guess that a person has dilated cardiomyopathy based on the symptoms. The best way to confirm the diagnosis is by taking an x-ray of the chest. The heart shows up as visibly enlarged in people with the disease.

If the disease is in its early stages, then doctors may not be absolutely sure that the enlarged heart is due to disease—the size of the heart can vary between people. In this case, the doctor may do an electrocardiogram to determine whether the heart is sending electrical signals properly. An echocardiogram can also distinguish dilated cardiomyopathy from other forms of heart disease. The map created of the heart can also highlight the enlarged chambers that occur at the beginning of the disease, before the heart muscle itself has enlarged to compensate for the larger interior volume.

Treatment

Doctors don't have any way of repairing damaged heart tissue. Furthermore, heart muscles don't have the equivalent of satellite cells to recognize the damaged muscle and repair it. Instead, once a heart has become weak due to dilated cardiomyopathy all the doctor can do is treat the symptoms of the disease and try to slow down how quickly the heart becomes even weaker. Despite attempts to treat dilated cardiomyopathy, about one third of people who are first diagnosed get temporarily better with treatment, one third stay about the same, and another one third continue getting worse.

The drugs that are used to treat dilated cardiomyopathy mainly help to make it easier for the heart to pump blood, or keep the heart beating regularly. With this added boost, there is less strain on the heart muscle itself and muscle fibers are less likely to break down. Other drugs prevent the blood from coagulating in the heart and sending blood clots throughout the body.

One type of drug that doctors use cause the arteries leaving the heart to relax and become larger. The heart does not have to work as hard to push blood through these larger arteries. Another group of drugs gets rid of excess fluid that builds up in a person's lungs. Other groups of drugs help the heart muscle fibers transmit the signal to contract, keeping the heart beating regularly.

Eventually, the heart gets so weak that it can't keep beating even with the help of drugs. At this point, the only treatment is to have a heart transplant. Despite all the people who need heart transplants, only about 2,000 hearts are donated each year. Many patients die while waiting for an appropriate heart. However, people who do receive a heart transplant have significantly better odds of living five years after their diagnosis than people who don't receive the new heart.

A recent device has made it more likely that people may live to receive their heart transplant. This device, called a left ventricular assistant device (LVAD), can take over almost all responsibility for pumping blood. Although these devices can't replace a heart, they do make it more likely that a person will survive the wait for a transplant.

RESTRICTIVE CARDIOMYOPATHY

Restrictive cardiomyopathy is the rarest of the three forms of cardiomyopathy. In this form of the disease, the heart muscle becomes extremely stiff and inflexible. The heart muscles cannot expand to let blood into the heart, then cannot squeeze to push blood out. This restricted movement led to the name restrictive cardiomyopathy.

Symptoms

People with restrictive cardiomyopathy cannot pump enough blood to the rest of the body. Although early in the disease the problem may only be noticeable when doing strenuous exercise, it slowly gets worse so that even day-to-day activities leave a person feeling out of breath.

Later in the disease, the heart becomes so inefficient at pumping blood that a person has heart failure and the usual symptoms of heart failure including congestion in the lungs. They may also feel nausea, bloating, or a loss of appetite from fluid building up around the liver and intestines.

Causes

Restrictive cardiomyopathy is not usually caused by a disease that directly affects the heart muscle. Instead, it is usually caused by diseases elsewhere in the body that eventually lead to symptoms in the heart. Among these diseases are hemochromatosis, in which iron builds up in the blood. Eventually, this iron is stored in other tissues including the heart and the liver, where it accumulates and prevents the cells from functioning normally.

In another disease that leads to restrictive cardiomyopathy, stringy proteins called amyloid build up in the heart muscle and form a tangled mass of protein that looks glassy under a microscope. Similarly, a group of syndromes called storage diseases can cause restrictive cardiomyopathy. These include Gaucher's disease and Fabry disease, in which a person is missing enzymes that break down molecules in the cell. These molecules accumulate in cellular structures called lysosomes.

Whether the cause is iron, amyloid, or other molecules, the heart fiber can't accommodate the added load and eventually becomes stiff and inflexible. Restrictive cardiomyopathy itself is not usually caused by inherited gene mutations, but the underlying disease, such as hemochromatosis or Gaucher's disease, is often inherited. However, there are a few small families that have inherited forms of restrictive cardiomyopathy without other underlying diseases. So far, however, there are too few families with these types of mutations for researchers to understand what genes cause the disease.

Diagnosis

Often a doctor can diagnose restrictive cardiomyopathy based on the symptoms or based on other diseases that a person has. For example, if a person has hemochromatosis and easily gets out of breath, a doctor can usually guess that the person has developed restrictive cardiomyopathy.

Unlike other forms of cardiomyopathy, this form cannot be detected easily by an echocardiogram. The heart is a normal size, and the muscle does not appear any thicker than normal heart muscle. Instead, the doctor usually has to take a heart muscle biopsy and look to see whether iron, amyloid, or other molecules have accumulated in the muscle fibers. The heart may also show characteristic patterns of how it transmits electricity in an EKG.

Treatment

Restricted cardiomyopathy tends to get worse with time. The drugs that help other forms of cardiomyopathy have not helped with this form of the disease. Instead, a person simply has to be careful not to stress the heart in order to maintain what heart function is remaining. Because the disease is often not diagnosed until the muscle is quite diseased, only about 30 percent of people who are diagnosed with restrictive cardiomyopathy live for five years after the diagnosis.

Acronyms

ADP Adenosine diphosphate

ASH Asymmetrical septal hypertrophy

ATP Adenosine triphosphate

CAD Coronary artery disease

DMPK Myotonic dystrophy protein kinase

DNA Deoxyribonucleic acid

DOMS Delayed onset muscle soreness

EKG Electrocardiogram

$FADH_2$ Flavine adenine dinuncleodite

FGF Fibroblast growth factor

FSHD Facioscapulohumeral dystrophy

HFL Harvard Fatigue Laboratory

HOCM Hypertrophic obstructive cardiomyopathy

IGF Insulin-like growth factor

IHSS Idiopathic hypertrophic subaortic stenosis

LVAD Left ventricular assistant device

MRF Myogenic regulator factors

NADHs Nicatinamide adenine dinuncleotide

NSAID Nonsteroidal anti-inflammatory drugs

OPMD Oculopharyngeal muscular dystrophy

PCR Polymerase chain reaction

RNA Ribonucleic acid

TGF-ß Transforming growth factor beta

Glossary

A band One of the stripes that are visible under a microscope in skeletal muscle. The A band is made up of myosin filaments in the center of the sarcomere.

Acetylcholine A molecule released by nerves that signals muscle fibers to contract.

Acetyl coenzyme A (acetyl CoA) A molecule that enters the citric acid cycle to produce energy. The acetyl CoA can come from sugars that have gone through glycolysis, or it can come directly from fats or proteins in the cell.

Actin One of the major proteins involved in muscle contraction. Actin proteins form a long fiber within the muscle contractile unit. Myosin, the other major protein involved in muscle contraction, attaches to the actin filaments and pulls the muscle shorter.

Acute muscle soreness The muscle soreness that happens immediately after intense exercise and quickly disappears. This soreness is often due to the acidity in muscles or from muscle swelling. It is distinct from delayed onset muscle soreness, which usually occurs the day after exercise.

Adenosine Triphosphate (ATP) A molecule that is used as a form of energy in the cell.

Aerobic exercise Exercise in which energy is made by processes involving oxygen. Types of aerobic exercise include swimming, biking, and jogging.

Anabolic steroids Hormones such as testosterone that cause muscles to grow larger. Some athletes take steroids to increase their muscle mass. Taking these drugs can have serious side effects and is banned in most sports.

Anaerobic exercise Exercise in which energy is made by processes that do not involve oxygen. Types of anaerobic exercise include weightlifting and sprinting.

Antagonistic pair Two muscles that have an opposite action, such as the muscle that bends the arm and the muscle that straightens the arm. The antagonistic pair controls and stabilizes the elbow as it bends and straightens.

Antibody A molecule made by the immune system that recognizes and binds to other molecules. Researchers can make antibodies that bind to proteins in muscles such as a specific type of myosin, and use that antibody to locate proteins within the muscle.

Atrium The upper chamber of the heart. Blood enters the atrium through veins leading in from the body, and leaves the atrium into the lower chamber of the heart, called the ventricles.

Autoimmune disease A disease in which the immune system attacks and destroys the body's own cells. Rheumatoid arthritis, multiple sclerosis, and some forms of inflammatory myopathy are all autoimmune diseases.

Autosome Chromosomes that do not play a role in determining a person's sex. Of the forty-six chromosomes in each human cell, forty-four are autosomes and two are sex chromosomes.

Becker muscular dystrophy The second most common form of muscular dystrophy. Children with the disease have muscles that become progressively weaker until they are not able to walk on their own. It is related to the most common form of the disease, Duchenne muscular dystrophy, but progresses more slowly.

Biopsy A technique in which a doctor removes a small part of a muscle for observation under a microscope.

Blood doping A technique used by some athletes to improve performance. The athlete draws blood several weeks before an event, then injects it soon before the event. The additional blood increases the amount of oxygen the blood can carry and improves performance. This technique is banned in most sports.

Calcium carbonate A molecule present in the muscle to help counter the acidity produced by intense anaerobic exercise. People who train for sprint-type events accumulate more calcium carbonate in their muscles to provide additional protection.

Calorie A measure of how much energy food contains.

Carbo-loading The process of eating a meal high in carbohydrates before an endurance event. When trained athletes carbo-load, their muscles store additional glycogen that the muscle can use for energy during the event.

Cardiac (heart) muscle The type of muscle found in the heart.

Cardiomyopathy A group of diseases that affect the heart muscle.

Carrier A person who carries one disease-causing mutation but does not have the disease. A carrier can pass the mutation on to his or her children. If the person's partner is also a carrier, then the child might inherit two mutated copies of the gene and could develop the disease.

Citric acid cycle A chemical reaction that takes place in the mitochondria. The cycle produces some energy for the cell and produces products that can be used to produce large amounts of energy through oxidative phosphorylation.

Collagen A tough, rope-like protein that makes up tendons and ligaments.

Concentric contraction The type of contraction that occurs when a muscle contracts and grows shorter, such as the biceps muscle when bending the elbow.

Congenital muscular dystrophy A form of muscular dystrophy that is evident in very young babies. Unlike other forms of muscular dystrophy, congenital forms of the disease do not get progressively worse with age.

Congestive heart failure One result of a heart that has become too weak to pump blood. Fluid builds up in the lungs, making a person seem congested.

Contracture A strong, sustained contraction that is common in people with some forms of muscular dystrophy. The contractures can be so strong that they cause bones to deform.

Creatine phosphate A molecule stored in the muscle that can quickly replenish ATP during a sudden burst of exercise.

Delayed onset muscle soreness The type of muscle soreness that occurs for several days after strenuous exercise. This type of soreness is due to muscle damage and is usually accompanied by muscle weakness.

Dilated cardiomyopathy The most common form of heart muscle disease. The heart chambers grow in size and the heart muscles becomes too weak to pump blood properly.

Distal myopathy A form of muscle disease in which the muscles farthest from the center of the body become progressively weaker.

Dominant mutation A mutation that causes disease when a person inherits a mutated copy of the gene from one parent and a normal copy of the gene from the other parent.

Duchenne muscular dystrophy The most common form of muscular dystrophy. Children with the disease have muscles that become progressively weaker starting in early childhood. Most children with the disease die while in their teens.

Dystrophin The protein that is mutated in Duchenne and Becker muscular dystrophies. Dystrophin is normally located on the outside of muscle fibers, where it seems to be involved in cell membrane function.

Eccentric contraction The type of contraction that occurs when a muscle contracts but the overall muscle grows longer rather than shorter. An eccentric contraction occurs in the biceps when lowering a heavy weight. The biceps muscle contracts to control the arm as it extends, but the muscle grows longer rather than shorter. Eccentric contraction is a common reason for muscle soreness after exercise.

Echocardiography A technique that gives doctors an image of the size and shape of the heart. This technique can help diagnose some forms of heart disease.

Ectoderm The outer cell layer in the developing embryo. These cells go on to form the skin, nerves, and eyes.

Electrocardiogram (EKG) A technique that tells doctors how the signal to contract is spread throughout the heart. This technique can help diagnose some forms of heart disease.

Electron transport chain A group of proteins embedded in the mitochondria. The electron transport chain takes electrons from the end products of the citric acid cycle and transports them to an oxygen molecule to create water. This process also produces ATP that the cell uses for energy.

Emery-Dreifuss muscular dystrophy A form of muscular dystrophy that causes muscles of the upper arms to become progressively weaker. People with this disease may also have diseased heart muscle.

Endoderm The inner cell layer of a developing embryo. These cells go on to form the digestive tract.

Facioscapulohumeral dystrophy (FSHD) A form of muscular dystrophy that affects the face, upper arms, and shoulders. It is a relatively mild form of muscular dystrophy that progresses slowly throughout a person's life.

Fascia The connective tissue surrounding an entire muscle. The fascia becomes part of the tendon at either end of the muscle, connecting the muscle to the bone.

Fascicle A bundle of muscle fibers within the muscle surrounded by a tissue called the perimysium. Each muscle is made up of many fascicles.

Fast-twitch A type of muscle fiber that is able to contract very quickly. Fast-twitch muscle fibers are predominantly found in muscles that must contract quickly and with great strength but do not need to contract over a long period of time. Sprinters tend to have more fast-twitch muscle fibers than endurance athletes.

Fibroblast growth factor (FGF) A protein that cells of the body use to signal developmental changes. In the developing muscle, FGF helps maintain immature muscle cells in a dividing state. Without FGF, cells divide too few times and produce smaller muscles.

Founder effect One explanation for why some mutations are more common in a particular population, such as the French Canadians or the Ashkenazi Jews. The people who founded those populations are thought to have carried some disease-causing mutations. Because people in those populations tended to marry amongst themselves, the mutation remained common although it might be quite rare in people worldwide.

Gene A small portion of a chromosome that carries the information to create a protein.

Gene chip A technology that allows researchers to identify which genes are being used by a tissue, organ, or group of cells.

Glucagon A hormone produced by the pancreas. Glucagon signals liver cells and fat cells to release fat and sugar into the bloodstream. Other cells, such as those of the muscle, use that fat and sugar to produce energy.

Glycogen A form of sugar stored in the liver and in the muscles. A muscle fiber can break down glycogen to make ATP for a contraction.

Glycolysis The process of breaking down glucose into two molecules of pyruvate. Glycolysis produces some energy for the cell and is the primary way of producing energy during anaerobic exercise. Pyruvate can either be recycled or converted into acetyl CoA for use in the citric acid cycle.

Gowers' maneuver A method of standing up that is characteristic of children with muscular dystrophy. The children push themselves into a crouching position, then walk the hands closer to the body pushing themselves upright. The children then use a nearby chair or other item to pull themselves up to a standing position.

H zone One of the stripes that are visible under a microscope in skeletal muscle. The H zone is the space between the two sets of actin filaments in the center of the sarcomere. The H zone grows smaller when the sarcomere contracts, and the actin filaments slide toward each other in the center of the sarcomere.

Heart failure The end result of some heart diseases, including hypertrophic cardiomyopathy. The heart becomes too weak to pump blood out to the body, even when resting.

Hypertrophic cardiomyopathy A form of heart disease in which a portion of the heart muscle becomes much larger than normal. The disease is usually found in the lower left-hand chamber of the heart, called the left ventricle. In severe forms of the disease, the enlarged heart muscle blocks blood from leaving the heart normally.

Hypertrophy The process in which muscles grow larger in response to exercise.

I band One of the stripes that are visible under a microscope in skeletal muscle. The I band is the region between the Z band at the outside of the sarcomere and the end of the myosin chain that spans the center of the sarcomere. As the muscle contracts and the Z disks move closer together, the I band grows smaller.

Inflammatory myopathy A form of muscle disease in which immune cells invade the muscles and cause them to grow progressively weaker.

Insertion The end of the muscle that is usually farthest from the center of the body. This end of the muscle is usually the one that moves when the muscle contracts. The end of the muscle closest to the body is called the origin.

Insulin A hormone produced by the pancreas after a meal. Insulin signals liver and muscle cells to take up sugar from the blood and store it as glycogen. Later, when less sugar is present in the blood, cells can use the stored glycogen to produce energy.

Insulin-like growth factor (IGF) A protein that seems to help the muscles maintain mass. Older people who are given IGF retain their muscle mass and improve their strength compared with other people in the same age group.

Intercalated disk A disk that separates two muscle fibers in the heart muscle. This disk can conduct the signal to contract from one muscle fiber to the next. With this connection, the entire heart muscle can contract in unison.

Isometric contraction The type of contraction that occurs when a muscle contracts but the joint doesn't open or close, such as when pushing against a wall or pushing down on a table.

Knock-out mouse A mouse in which researchers have completely eliminated the function of one gene. Knock-out mice are commonly used to help researchers understand the function of individual genes.

Lactate threshold The point at which muscles begin accumulating lactic acid as a result of intense exercise. When athletes pass their lactate threshold, they are not as able to sustain the same intensity in exercise. Many people consider an athlete's lactate threshold to be an indicator of fitness.

Lactic acid A molecule that can be produced as the end product of glycolysis. Glycolysis produces a molecule called pyruvate that can either enter the citric acid cycle when enough oxygen is present in the cell or be converted to lactic acid. Lactic acid accumulates in a muscle cell during anaerobic exercise, causing the muscle to become acidic and painful.

Limb girdle muscular dystrophy A form of muscular dystrophy in which muscles of the stomach and back become progressively weaker. Symptoms can begin at any time but are usually apparent by the time a person is in his or her 20s.

Mesoderm The middle cell layer in a developing embryo. These cells go on to form the organs and muscles.

Microtear Small tears in the proteins that make up the tendons or ligaments. Microtears can be painful and also make the structure weaker and more susceptible to further injury.

Mitochondria A unit within the cell that generates ATP for contraction. Some ATP can be made through reactions that take place outside the mitochondria, but the majority of ATP for contraction comes from processes with the mitochondria.

Motor unit A group of muscle fibers that all contract in response to the same nerve signal.

Muscle fiber A muscle unit made up of many muscle cells that have fused together. Each fiber receives the signal to contract from a single nerve. The muscle fiber is surrounded by a cell membrane called the sarcolemma.

Muscular dystrophy A general term used to describe muscle diseases in which the muscle becomes progressively weaker. Some forms of muscular dystrophy begin in childhood and cause an early death, while other forms progress more slowly and are less disabling.

Mutation An alteration within a gene. Some mutations are harmless, while others can prevent the gene from producing a normal protein. Many diseases are caused by mutations.

Myoblast A developing muscle cell. A cell becomes a myoblast once it receives a signal to develop as a muscle early in development.

MyoD A protein that is made by muscle cells. This protein is characteristic of muscle cells and can be used to distinguish those cells in the embryo that are developing into muscles.

Myofibril The contractile unit within a muscle fiber. The myofibril is made up of a series of contractile units called sarcomeres. Each muscle fiber contains many myofibrils, all of which contract when the muscle fiber receives a signal from a nerve.

Myogenic regulator factor (MRF) Proteins that are made by developing muscle cells and that direct the cell through the phases of development. These proteins direct the cell to migrate to the limbs, then control the cell as it matures into a muscle cell.

Myoglobin A molecule in the muscle that collects oxygen from the blood and delivers it to mitochondria in the muscle fiber.

Myopathy Any disease of the muscle.

Myosin One of the major contractile proteins making up a muscle fiber. Myosin proteins form chains that pull on actin filaments, causing the muscle fiber to contract.

Myotonic dystrophy A group of diseases in which the muscles are slow to relax after contracting. The disease primarily affects skeletal muscle but can also affect smooth and heart muscle.

Neural tube A primitive spinal column. During embryonic development, the neural tube releases proteins that direct some nearby cells to begin developing into immature muscle cells.

Nonsteroidal anti-inflammatory drugs (NSAIDs) A class of drugs that reduce inflammation and pain. The most common NSAID is ibuprofen.

Notochord A primitive support structure that runs down the back of an embryo. During embryonic development, the notochord releases proteins that direct some nearby cells to begin developing into immature muscle cells.

Oculopharyngeal muscular dystrophy (OPMD) A form of muscular dystrophy that first becomes apparent in the muscles of the eyes and throat. The disease usually begins when a person is in his or her 50s.

Origin The end of the muscle that is closest to the center of the body. The origin is usually the end that remains in place when a muscle contracts. The other end, called the insertion, is usually attached to a bone that moves during a contraction.

Oxidative phosphorylation The process of combining electrons with oxygen to create water. This process also produces energy for the cell when enough oxygen is present. It is the main process used to generate energy during aerobic exercise.

Pacemaker A nerve bundle within the heart that signals the heart to beat at a regular rhythm. If the natural pacemaker fails to function normally, an artificial pacemaker can be inserted to keep the heart beating regularly.

Pax-3 A protein made by developing muscle cells that delays those cells from maturing into muscles. This protein prevents the immature muscle cells from maturing before they migrate to the appropriate location in the limbs. Mice that do not make pax-3 end up with muscle cells that mature while in the process of migrating and fail to form muscles in the appropriate location.

Perimysium Connective tissue that surrounds the bundle of muscle fibers making up a fascicle.

Polymerase chain reaction (PCR) A process used to amplify a gene sequence into many copies. PCR provides enough DNA to carry out further experiments.

Primary myoblasts Immature muscle cells that migrate to the developing limbs. The primary myoblasts establish the location of future muscles. Secondary myoblasts migrate after the primary myoblasts and add volume to the muscles established by the primary myoblasts.

Pyruvate The end product of glycolysis. Pyruvate can be either converted into acetyl CoA for use in the citric acid cycle, or converted into lactic acid during anaerobic exercise.

Recessive mutation A mutation that causes disease only when a person inherits the mutation from both parents. If a person only inherits one copy of a recessive mutation, he or she does not develop the disease, although it can be passed on to his or her children.

Restrictive cardiomyopathy A form of heart muscle disease in which the heart becomes inflexible and unable to pump blood.

Sarcolemma The membrane that surrounds a muscle fiber. It is similar to the cell membrane of other cells in the body.

Sarcomere An individual contractile unit within the myofibril. The sarcomere contains actin filaments attached to either end and myosin chains that pull the actin filaments closer together, making the sarcomere grow shorter. When all the sarcomeres in a myofiber contract, the entire fiber shortens.

Sarcoplasmic reticulum A network of tubules that runs throughout the muscle fiber. The sarcoplasmic reticulum stores calcium when the fiber is not contracted and releases calcium when the fiber receives a signal to contract.

Satellite cells Immature muscle cells found throughout a muscle. These cells can mature into fast-twitch or slow-twitch muscle fibers if those fibers become damaged. They are the muscles' primary way of repairing injury.

Secondary myoblasts Immature muscle cells that form the second wave of cells migrating to the limb. They attach to muscle scaffolding formed by the primary myoblasts.

Septum The wall that divides the left and right chambers of the heart. In hypertrophic cardiomyopathy, the septum may grow so large that blood cannot leave the heart.

Sex-linked inheritance A form of inheritance in which one sex is more likely to inherit the trait. Most sex-linked traits, such as muscular dystrophy, are much more common in males than in females.

Simple twitch A single contraction of a muscle fiber in response to an isolated nerve signal. Most contractions involve a sustained nerve signal that causes the muscle to remain contracted for a longer period of time.

Skeletal muscle Muscles that are attached to the skeleton and allow the body to move. This is the most common type of muscle in the body. It is also called voluntary muscle because these are the muscles that move voluntarily, or striated muscle because of the stripes, or striations, that can be seen in the muscle under a microscope.

Skinned fibers Muscle fibers that have had the outer membrane removed. Researchers can expose the skinned fibers to different chemicals and look for changes in how the fiber contracts.

Slow-twitch A type of muscle fiber that is able to contract very quickly. Slow-twitch muscle fibers are predominantly found in muscles that must contract repeatedly but without much strength. Endurance athletes tend to have more slow-twitch muscle fibers than sprinters do.

Smooth muscle The type of muscle that surrounds organs and tissues such as the bladder, the blood vessels, or the iris of the eye. These muscles are not under voluntary control and are therefore sometimes called involuntary muscles.

Somite Paired structures running along the back of a developing embryo. Cells that make up the somites go on to form muscles, nerves, and other structures.

Sphincter A skeletal muscle that forms a circular band. Sphincter muscles usually control the size of an opening, such as the mouth or the entrance to the stomach. The muscle contracts to close the opening or relaxes to open it.

Sprain A type of injury that occurs when the ligaments get pulled or torn. The ligaments around the ankle joint are the most commonly sprained.

Strain A type of injury that occurs when a muscle or tendon gets torn. Some strains are relatively minor, while others can cause long-term muscle damage.

Sudden death An unexpected death that occurs when the heart suddenly stops functioning. People with hypertrophic cardiomyopathy are at high risk of sudden death.

Synergist A muscle that works in conjunction with an antagonistic pair to control the movement of a joint. The synergist usually runs beside a joint or diagonally across a joint. It does not act on its own to open or close the joint, but does control and stabilize the contraction of other muscles.

T Tubule Tubules that run through muscle fibers carrying the signal to contract. The signal passes from the T tubule to the sarcoplasmic reticulum, which releases calcium and causes the contraction to take place.

Tendon A band of connective tissue that connects the muscle to the bone.

Testosterone A hormone that produces male characteristics including large muscles.

Tetanus contraction A sustained contraction as a result of many independent signals from a nerve.

Transforming Growth Factor (TGF-ß) A protein that cells of the body use to signal developmental changes. In the developing muscle, TGF-ß is produced by muscle cells that have fused into muscle fibers. The TGF-ß signals nearby myoblasts not to fuse with the growing muscle and instead to remain as immature muscle cells. This signal helps regulate the eventual size of the muscle.

Tropomyosin A protein that forms long filaments wrapping around actin within the muscle fiber.

Troponin A protein that is associated with actin and tropomyosin within the muscle fiber.

Ventricle The lower chamber of the heart. Blood enters the ventricle from the upper heart chamber, called the atrium, and leaves the ventricle through arteries leading out to the body.

VO_2max A measurement of how much oxygen a person can use to produce energy. Highly trained athletes can use more oxygen than untrained people, providing their cells with more energy and improving athletic performance.

Z band One of the stripes that are visible under a microscope in skeletal muscle. The Z band is a dense area that separates the sarcomeres. The actin filaments are embedded in the Z band, extending inward into each sarcomere.

Organizations and Web Sites

Cardiomyopathy Association
40 The Metro Centre
Tolpits Lane
Watford, Hertz WD18 9SB
UK
Phone: + 44 (0) 1923 249 977
Fax: +44 (0) 1923 249 987
Email: info@cardiomyopathy.org
www.cardiomyopathy.org/homepage.htm

This site provides information about cardiomyopathy for both patients and doctors. It also contains news about ongoing research in the cardiomyopathies.

Facioscapulohumeral Muscular Dystrophy Society
3 Westwood Road
Lexington, MA 02420
Phone: (781) 860-0501
Fax: (781) 860-0599
www.fshsociety.org

This site provides information about Facioscapulohumeral Muscular Dystrophy (FSHMD) and contains links to the society's publication *FSH Watch*.

Hypertrophic Cardiomyopathy Association
www.4hcm.org

With information about hypotrophic cardiomyopathy symptoms, biology, and treatment, this site is a comprehensive overview of the disease. It also includes a link to the association's newsletter and information about clinical trials.

International Myotonic Dystrophy Organization
P.O. Box 1121
Sunland, CA 91041-1121
Phone: (866) 679-7954
Email: info@myotonicdystrophy.org
www.myotonicdystrophy.org

This web forum provides support and information for parents taking care of a child with myotonic dystrophy.

Logan Paige Foundation for Myotonic Dystrophy
Email: info@loganpaige.org
www.loganpaige.org/index1.html

This foundation web page contains links to other organizations, medical articles, and news outlets with information relating to myotonic dystrophy.

Muscular Dystrophy Association
3300 East Sunrise Drive
Tucson, AZ 85718
Phone: (800) 572-1717
Email: mda@mdausa.org
www.mdausa.org

This site has news and information about muscular dystrophy, links to the association's newsletter, and a forum where patients and family members can have questions answered by an expert.

Muscular Dystrophy Family Foundation
2330 North Meridian Street
Indianapolis, IN 46208-5730
Phone: (317) 923-6333 or (800) 544-1213
www.mdff.org

This site provides support to families and patients with all forms of muscular dystrophy. The site is focussed on helping kids live independent lives.

Parent Project Muscular Dystrophy
1012 North University Boulevard
Middletown, OH 45042
Phone: (513) 424-0696 or (800) 714-5437
Fax: (513) 425-9907
www.parentprojectmd.org

This site provides extensive news about research strategies for Duchenne and Becker muscular dystrophies. It also has background information about the disease and links to legislation that affects families and children with these diseases.

Bibliography

Almekinders, L. C. "Anti-inflammatory Treatment of Muscular Injuries in Sport: An Update of Recent Studies." *Sports Medicine* 28, no. 6 (1999): 383–388.

Amelink G. J., W. A. van de Wal, et al. "Exercise-induced Muscle Damage in the Rat: The Effect of Vitamin E Deficiency." *Pflugers Archives* 419, no. 3–4 (October 1991): 304–309.

Avery, G., M. Chow, and H. Holtzer. "An Experimental Analysis of the Development of the Spinal Column V. Reactivity of Chick Somites." *Journal of Experimental Zoology* 132 (1956): 409–425.

Buckingham M., L. Bajard, et al. "The Formation of Skeletal Muscle: From Somite to Limb." *Journal of Anatomy* 202, no. 1 (January 2003): 59–68.

Byrne, C., C. Twist, and R. Eston. "Neuromuscular Function After Exercise-induced Muscle Damage: Theoretical and Applied Implications." *Sports Medicine* 34, no. 1 (2004): 49–69.

Campbell, N. A. *Biology*, 4th ed. Menlo Park, CA: Benjamin/Cummings, 1997.

Carlson, B. M., and J. A. Faulkner. "The Regeneration of Skelatal Muscle Fibers Following Injury: A Review." *Medicine and Science in Sports and Exercise* 15, no. 3 (1983): 187–198.

Ebbeling, C. B., and P. M. Clarkson. "Exercise-induced Muscle Damage and Adaptation." *Sports Medicine* 7 (1989): 207–234.

Eggleton, P., and G. P. Eggleton. *Journal of Physiology* 63 (1927): 155–161.

Emerson, C. P., and H. L. Sweeney. *Methods in Muscle Biology*. San Diego, CA: Academic Press, 1997.

Emery, A. E. *Duchenne Muscular Dystrophy*, 2nd ed. Oxford and New York: Oxford University Press, 1993.

Factor, P. *Gene Therapy for Acute and Acquired Diseases*. Boston: Kluwer Academic Publishers, 2001.

Ferrari, R. *Hypertrophic Cardiomyopathy: From Molecular and Genetic Mechanisms to Clinical Management*. London: Saunders, 2001.

Fulton, J. F. *The History of the Physiology of Muscle 1899–1930*. Baltimore: Williams & Wilkins Co., 1930.

Garrett, W. E., Jr. "Muscle Strain Injuries." *American Journal of Sports Medicine* 24 (1996): S2–S8.

Gilbert, S. F., M. S. Tyler, and R. N. Kozlowski. *Developmental Biology*, 6th ed. Sunderland, MA: Sinauer Associates, 2000.

Griggs, R. C., and G. Karpati. *Myoblast Transfer Therapy.* New York: Plenum Press, 1990.

Gussoni, E., G. K. Pavlath, A. M. Lanctot, et al. "Normal Dystrophin Transcripts Detected in Duchenne Muscular Dystrophy Patients After Myoblast Transplantation." *Nature* 356 (1992): 435–438.

Hacein-Bey-Abina, S., C. Von Kalle, M. Schmidt, et al. "LMO2-Associated Clonal T Cell Proliferation in Two Patients After Gene Therapy for SCID-X1." *Science* 302 (October 17, 2003): 415–419.

Huxley, Andrew. *Reflections on Muscle.* Princeton, NJ: Princeton University Press, 1980.

Jennison, S. H., and B. A. Mulch. "Cardiomyopathy and Cardiac Muscle Disease." *Primary Care Management of Heart Disease* 24 (2000): 288–295.

Johansson, P. H., L. Lindstrom, G. Sundelin, and B. Lindstrom. "The Effects of Preexercise Stretching on Muscular Soreness, Tenderness and Force Loss Following Heavy Eccentric Exercise." *Scandinavian Journal of Medicine and Science in Sports* 9, no. 4 (August 1999): 219–225.

Lieber, Richard L. *Skeletal Muscle Structure and Function: Implications for Rehabilitation and Sports Medicine.* Baltimore: Williams & Wilkins, 1992.

McArdle, William D., Frank I. Katch, and Victor L. Katch. *Exercise Physiology: Energy, Nutrition, and Human Performance*, 5th ed. Philadelphia: Lippincott Williams & Wilkins, 2001.

Phillips, Chandler A., and Jarold S. Petrofsky. *Mechanics of Skeletal and Cardiac Muscle.* Springfield, IL: Thomas, 1983.

Reid, T., R. Warren, and D. Kim. "Intravascular Adenoviral Agents in Cancer Patients: Lessons from Clinical Trials." *Cancer Gene Therapy* 9 (2002): 979–986.

Sarnat, H. B. "Muscular dystrophies." In *Nelson Textbook of Pediatrics*, 15th ed. Edited by Richard E. Behrman, Robert M. Kliegman, Ann M. Arvin, et al. Philadelphia: W. B. Saunders, 1996.

Soloman, Eldra P., Linda R. Berg, Diana W. Martin, et al. *Biology*, 4th ed. Orlando, FL: Harcourt Brace & Company, 1997.

Strohman, Richard C., and Stewart Wolf. *Gene Expression in Muscle.* New York: Plenum Press, 1985.

Sugi, Harno. *Current Methods in Muscle Physiology: Advantages, Problems, and Limitations.* New York: Oxford University Press, 1998.

Tiidus, P. M., and E. Bombardier. "Oestrogen Attenuates Post-exercise Myeloperoxidase Activity in Skeletal Muscle of Male Rats." *Acta Phyiologica Scandinavica* 166, no. 2 (June 1999): 85–90.

Wilmore, Jack H., and David L. Costill. *Physiology of Sport and Exercise*, 2nd ed. Champaign, IL: Human Kinetics Publishers, 1999.

Index

About the Author

AMY ADAMS is an independent scholar who has written for *The Scientist*, Discovery Channel Online, *CBS Health Watch*, and *Science* among others.